Peri-Urban China

The urban–rural relationship in China is key to a sustainable global future. This book is particularly interested in peri-urbanization in China, the process by which fringe areas of cities develop.

Recent institutional change has helped clarify property rights over collective land, facilitating peri-urban area development. Chapters in this book explore how rural industrialization has changed the landscape and rules about land use in peri-urban areas. It looks at the role of rural industrialization and provides a detailed exploration of peri-urbanization theory, policy, and its evolution in China. Leading discussions find out how fragmented bottom-up industrialization, urbanization, and lax governance have led to a series of social and environmental problems. The progress in redevelopment of peri-urban areas was initially slow due to the spatial lock-in effect. This book offers practical solutions to environmental issues and explains how policymakers have the potential to redevelop a future collaborative, inclusive, and sustainable approach to peri-urban areas.

This in-depth approach to urbanization will be useful to academics in urban planning and governmental organizations. It will also be advantageous to NGOs and professionals involved in urban planning, public administration, as well as land-use work in China and other developing countries.

Li Tian is a professor in Urban Planning at Tsinghua University, China.

Yan Guo is an associate professor in Urban Planning at Wuhan University, China

Routledge Critical Studies in Urbanism and the City

This series offers a forum for cutting-edge and original research that explores different aspects of the city. Titles within this series critically engage with, question, and challenge contemporary theory and concepts to extend current debates and pave the way for new critical perspectives on the city. This series explores a range of social, political, economic, cultural, and spatial concepts, offering innovative and vibrant contributions, international perspectives, and interdisciplinary engagements with the city from across the social sciences and humanities.

Art and the City
Worlding the Discussion through a Critical Artscape
Edited by Julie Ren and Jason Luger

Gentrification as a Global Strategy
Neil Smith and Beyond
Edited by Abel Albet and NúriaBenach

Gender and Gentrification
Winifred Curran

Socially Engaged Art and the Neoliberal City
Cecilie Sachs Olsen

Peri-Urban China
Land Use, Growth, and Integrated Urban–Rural Development
Li Tian and Yan Guo

Cities and Dialogue
The Public Life of Knowledge
Jamie O'Brien

For more information about this series, please visitwww.routledge.com/ Routledge-Critical-Studies-in-Urbanism-and-the-City/book-series/RSCUC

Peri-Urban China

Land Use, Growth, and Integrated Urban–Rural Development

Li Tian and Yan Guo

Routledge
Taylor & Francis Group
LONDON AND NEW YORK

Figures

Tables

Acknowledgements

This book is supported by National Key R&D Program of China (Project number: 2018YFD1100105), the Outstanding Young Scientist Program of Beijing, and the National Science Fund of China (Project number: 41601153, 51878367). The authors wish to express gratitude to Professor Richard LeGates from San Francisco State University and Professor Jieming Zhu from Tongji University for their constructive comments.

Dr. Lin Zhou, Ms. Yaqi Yan, Ms. Yaxin Wu, and Ms. Jingwei Li from Tsinghua University and Dr. Zhihao Yao from Tongji University participated in data, graphics preparation and processing, and writing of some parts of the manuscript, and they deserve special thanks for their contribution to this book.

We are grateful for the support of the editors and staff at Routledge who enthusiastically supported this project and helped see the manuscript through to completion. We are particularly indebted to Ruth Anderson who managed this book. Special thanks also go to Brian Humek for his help in proof-reading and writing.

1 Introduction

While urbanization has been a global phenomenon, its pattern and forms are not universally uniform. Some countries might not follow the same path as others. Peri-urbanization is part of the uneven urban landscape, and the differences when urban and rural become entangled in the process of peri-urbanization. In Southeast Asia, peri-urban areas are known as extended metropolitan regions (Ginsburg, 1990), Desakotas (McGee, 1991), and ruralo-polises (Qadeer, 2000). Peri-urbanization is characterized by rapid changes in land use, building styles, economic activities, inconsistencies between administrative structures and territory, and influxes of new population. Visually, peri-urbanization blurs the preconceived distinct boundaries between rural and urban. Meanwhile, peri-urbanization is nearly always associated with problems, for example, lack of planning and infrastructure, degradation of environment, or loss of agricultural land. Therefore, peri-urbanization can be understood as an attempt to make sense of change, and create a new category for a phenomenon that does not fit existing categories (Willis, 2005).

With the rapid industrialization in rural China, the traditional countryside landscape has been gradually replaced by multi-storey residential blocks and scattered factory buildings over the past several decades (Tian, 2015). Given the vastness of Chinese territory, socio-economic situation and institutional arrangement vary from region to region. The peri-urban regions first emerged in developed areas such as the Pearl River Delta (PRD) and Yangtze River Delta (YRD), followed by Beijing–Tianjin–Hebei Region (BTH). In the southwest region, such as Chengdu-Chongqing region, and the Fujian province of south region, peri-urban areas also emerged. In the less-developed areas such as Northeastern China, however, urban–rural boundary is clearly defined and there have been no typical peri-urban regions (Figure 1.1).

Land-use transformations are generated by fundamental economic and social changes, and the difference of development model has resulted in diversified land-use pattern and landscape. In the PRD region, village-based land development has facilitated the industrialization process and economic growth in the peri-urban area; however, it is also one of the main reasons leading to urban sprawl and land fragmentation in peri-urban areas. In the YRD region, township-based development has consolidated land use to

Figure 1.1 Spatial distribution of peri-urban areas in China
Source: drawn by authors

some extent and land use is more consolidated in the peri-urban area. In either case, however, the land-use dynamics has exhibited the dual characteristics of fragmentation and aggregation. The comparison between various peri-urban areas indicates the need to emphasize the combination of top-down and bottom-up approaches in the interest of sustainable land use in the peri-urban areas of China.

There has been a wealth of literature documenting the urbanization of China, and its social, economic transformation, and challenges since the reform opening (Ren, 2013; Zhang et al., 2016; Eggleston et al., 2017). However, research on peri-urban areas with dynamic manufacturing-driven economy and fragmented urbanization has been scarce. In reality, the future direction of these areas is essential for sustainable and compact urbanization in China. Through extensive and intensive case studies, this book outlines a picture of growth and land use in the peri-urban areas of three of China's most developed urban agglomerations: the YRD, the PRD and the BTH region. The book begins with literature review on the urban–rural relationship and examines the socio-economic development path of Chinese peri-urban areas. It then reviews the evolution of rural industrialization and analyzes the land-use dynamics of peri-urban areas in the three urban agglomerations. From the institutional perspective, it examines the role of village collectives and villagers in capturing land surplus value and the consequent land fragmentation. Moreover, the governance structure and the recent local practice of redevelopment in peri-urban areas are analyzed. This book concludes with policy implications to make urbanization compact and equal in the peri-urban areas of China.

1.1 Research questions

In this book, we mainly address the following three research questions:

(1) What are the spatio-temporal characteristics of land-use change in typical peri-urban areas of China?
(2) What is the social and economic driving force behind land-use dynamics of peri-urban areas of China?
(3) What is the policy implication of sustainable land development in peri-urban areas?

1.2 Research method

In this research, we first adopt case-study approach to present the land-use dynamics of peri-urban area. The development pattern is very diverse in peri-urban areas of China, and it is difficult to outline an overall picture of peri-urban areas. Therefore, we select typical peri-urban areas in three agglomerations of China: Jiangyin and Kunshan in YRD, Nanhai and Shunde in PRD, and Shunyi in BTH to conduct research of land-use change in peri-urban areas.

Moreover, we combine qualitative and quantitative analyses, and apply multi-disciplinary tools, including GIS, landscape ecology indices, and urban planning tools to reveal the land-use characteristics of peri-urban areas. In terms of a theoretical framework, this book mainly adopts the New Institutional Economics (NIE) and property rights theories to analyze the growth and land use of peri-urban areas. It examines the role of institutional arrangements and governance in shaping land use from the perspective of property rights and transaction cost, and explains why the land use is locked-in under the existing institutional arrangement.

1.3 The structure of the book

This book has been organized as follows:

1.3.1 Chapter 1: Introduction

The introduction explains the research context, significance, objectives, and research methods of this research, and briefly discusses the framework of this book.

1.3.2 Chapter 2: Theories and Debates of Urban–Rural Development

The aim of this literature review is to examine the complexity of the theoretical discussion on concepts and models of peri-urban areas within the broader literature on rural–urban interactions and linkages. This literature review begins with early 20th-century debate, and goes through different models and theories of urban/rural bias. More recent literature is examined, basically from the 1990s to present day.

1) Classical economic theories of urban–rural development
2) Urban bias vs rural bias
3) Urban–rural linkages and flows
4) Globalization and extended metropolitan regions
5) Peri-urbanization: concepts, models and spatial patterns

1.3.3 Chapter 3: Evolution of Urban–rural Relationship and Peri-urban Areas Development in China

This chapter goes through the evolution of the urban–rural relationship since the establishment of the People's Republic of China in 1949 in order to provide a backdrop to understand "dual-track" urbanization: state-led and bottom-up urbanization. We select the peri-urban areas of Shanghai in the YRD, Guangzhou in the PRD, and Beijing in the BTH as case studies, examine their social-economic changes since the 1990s, and analyze the socio-economic characteristics of peri-urban areas.

1) An overview of urban–rural relationship in China
2) Urban–rural divide from 1949 to 1978
3) Urban–rural interaction since the reform opening
4) Socio-economic development of peri-urban areas: case studies of peri-urban areas in three urban agglomerations

1.3.4 Chapter 4: Industrialization, Fragmented Peri-urbanization, and Land-Use Dynamics

This chapter reviews the rural industrialization process driven by Township–village-enterprises (TVEs) and the burgeoning industrial clusters dominated by urban states after the demise of TVEs, and the various forms of city expansion in the social-economic transition context. Rapid urbanization in peri-urban areas has been driven by both the spill-over effects of the central city and non-agricultural land growth led by numerous autonomous rural collectives, resulting in a fragmented urbanizing landscape. By analyzing the land-use dynamics of typical peri-urban areas, including Jiangyin and Kunshan in YRD, Nanhai and Shunde in PRD, and Shunyi in BTH, this chapter identifies their spatio-temporal characteristics of land-use change such as land fragmentation, mix of residential and industrial land, and fragmented property rights since the 2000s.

1) Industrialization facilitated by TVEs and industrial clusters in peri-urban areas
2) Peri-urbanization characterized by fragmented land use and governance
3) Spatio-temporal characteristics of land-use dynamics in three peri-urban areas
4) What drives land-use dynamics of peri-urban areas in three peri-urban areas?

1.3.5 Chapter 5: Land Development under Institutional Uncertainty and Land Rent Seeking in Peri-urban Areas

Urban China and rural China are two institutionally distinctive domains, and rights over collective owned land are ambiguously delineated. The dual land management system has created institutional uncertainty, and effective state governance is absent in peri-urban areas. This chapter first analyzes the nature of ambiguity of the property rights over collectively owned land, followed by an analysis of formal or informal institutional change given the ambiguity of collective land. The characteristic of institutional uncertainty and how the institutional settings affect the modes of land development and land rents competition are then discussed. Finally, the results of land development under different ways of land rents distribution are analyzed and compared among cases in different regions.

1) Characteristics of China's transitional institutions.
2) The change of land-related institutions and land markets.
3) Landed interests formation and their interaction during peri-urban development.
4) Land development and land rents distribution in Nanhai: a case of PRD.
5) Land development and land rents distribution in Kunshan: a case of YRD.

1.3.6 Chapter 6: Institutional Change to Redevelop Peri-urban Areas and Spatial Lock-in

In 2009, the state launched the "Three-olds Renewal" institutional reform in Guangdong province. With this reform, the state delegated certain powers to existing land users in order to facilitate the redevelopment of dilapidated urban areas, old villages, and old factories and to formalize the informal development rights over collective land in the peri-urban areas. Applying the framework of institutional change, this chapter examines the new arrangement of redevelopment policies, looking closely at the role of the state and power relations. By taking the typical peri-urban areas, Nanhai and Panyu in the PRD, as cases, this chapter presents local practices in which institutional changes are made in the redistribution of benefits between the state, village collective, and villagers. Spatial lock-in effect due to the reliance of local cadres on landed interests and high transaction costs to achieve consensus for redevelopment, however, has made redevelopment of peri-urban areas a time-consuming and complicated process.

1) Three-olds renewal: policy and process
2) Constraints over three-olds renewal: multiple perspectives
3) Three-olds renewal in Nanhai: characteristics and cases of projects
4) Three-olds renewal in Panyu: characteristics and spatial lock-in

1.3.7 Chapter 7: Conclusion: Towards compact and integrated urban -rural development in China

This chapter summarizes the findings from the case study and explores the policy implications to achieve the goals of sustainable and integrated urban–rural development in the peri-urban areas as seen from the perspectives of economic development, spatial change, and governance change.

1) Transition from village-initiated industrialization to township/municipality coordinated development
2) Transition of spatial growth pattern from rural fragmentation to compact urbanization
3) Transition of governance model from rural autonomy to urban–rural integrated governance

References

Eggleston, K., Oi, J. C., &Wang, Y. (2017).*Challenges in the Process of China's Urbanization*. Palo Alto: Shorenstein Asia-Pacific Research Center.

Ginsburg, N. (1990). Extended Metropolitan Regions in Asia: A New Spatial Paradigm, In Ginsburg, N. (Ed.) *The Urban Transition: Reflections on American and Asian Experiences*, Hong Kong: Chinese University of Hong Kong Press, 21–42.

McGee, T. G. (1991). The Emergence of Desakota Regions in Asia: Expanding a Hypothesis, In Ginsburg, N., Koppel, B., &McGee, T. G. (Eds.) *The Extended Metropolis: Settlement Transition in Asia*, Honolulu: University of Hawaii Press, 3–25.

Qadeer, M. A. (2000).Ruralopolises: The Spatial Organization and Residential Land Economy of High-Density Rural Regions in South Asia. *Urban Studies*, 37 (09):1583–1603.

Ren, X. F. (2013).*Urban China*. Cambridge: Polity Press.

Tian, L. (2015).Land Use Dynamics Driven by Rural Industrialization and Local Land Finance in the Peri-urban Areas of China: The Examples of Jiangyin and Shunde. *Land Use Policy*, 45: 117–127.

Willis, A. M. (2005). From Peri-urban to Unknown Territory. Paper presented at the State of Australian Cities National Conference.

Zhang, Q., Sun, Z., Wu, F., &Deng, X. (2016).Understanding Rural Restructuring in China: The Impact of Changes in Labor and Capital Productivity on Domestic Agricultural Production and Trade. *Journal of Rural Studies*, 47:552–562.

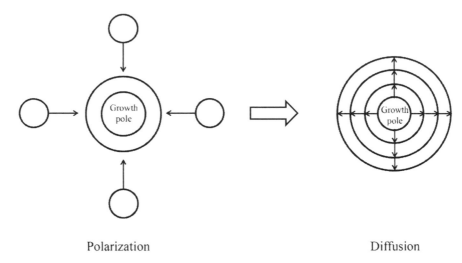

Polarization   Diffusion

Figure 2.1 Polarization-diffusion effect in the growth pole theory.
Source: drawn by authors

framework of uneven development in the center–periphery relationship, which stresses that regional imbalances based on chronological gaps appear inherent in the integration processes (Dobrescua & Dobre, 2014). The center–periphery relationship explains the structural relationship between the advanced or metropolitan "center" and a less developed "periphery," either within a particular country, or between capitalist and developing societies.

The theories of the growth pole and uneven development have been associated with a centralized "top-down planning" system (Stöhr, 1981). These theories were highly accepted by the centralized and "monolithic" planning system in most developed countries. For instance, since the establishment of the People's Republic of China, the central government has adopted an industrialization-oriented and urban-biased institutional arrangement, and much less resources have been delivered to the rural area except during the cultural revolution when the state policy was diverted from urban areas to the countryside, in part, due to the collective endeavor of communes (Nolan & White, 1984). Nevertheless, the urban bias polices were readopted after the reform opening. With the introduction of the tax sharing system reform in 1994, local government's preference for urban bias policies was enhanced, leading to increasing urban–rural income disparity and great social risks. Moreover, experiences from Latin America and Africa have proved that the urban bias strategy has found difficulty in achieving success. In contrast, it has maintained or even increased inequality between urban and rural areas, especially since the expected "trickle down effects" have been replaced instead by adverse "backwash effects," resulting in a rural–urban conceptual dichotomy (Unwin, 1989).

Policies formulated from this optimistic conception emphasize the role of the city, industrial orientation, and the development of capital-intensive and high-tech industries, showing the theory has an obvious urban bias. Due to the urgent pursuit of economic growth in developing countries, although some authors show evidence demonstrating its lack of success, *growth pole–* oriented policies have been widely used in developing countries (Adell, 1999).

2.2.2 Urban bias critique

In the 1970s, the traditional top-down theory of urban and rural development was seriously challenged, causing a fierce criticism of the urban bias development. Perhaps, the most influential and strongest critique of uneven development generated or maintained by urban-oriented policies in poor countries came from Michael Lipton (1977). The conflict between rural classes and urban classes became the most important class conflict in poor countries. Lipton's basic argument of "urban bias" (1977) was that urban dwellers, having far more power than those in rural areas, were able to divert a disproportionate share of resources toward their own interests and against the rural sector. This flowing of resources was detrimental to the development of rural areas, making the poor poorer and causing inequality within rural areas.

Bates (1981) further developed several of Lipton's ideas, mainly regarding the political force of urban pressure groups. They both posited that "governments concerned with political survival must accommodate the economic interests of urban workers, managers and civil servants" (Lofchie, 1997). Unwin (1989) pointed out that in assuming the main difference between rural and urban consists of their social "classes," and the drawback of Lipton's urban bias theory was the conflation of people and space. Most scholars have accepted urban bias in development policy, basically related to a preference for import substitution industrialization policies being maintained (Adell, 1999).

The proposal of "urban bias" theory has triggered the exploration of bottom-up development strategy. For instance, Kenya and Côte d'Ivoire adopted rural bias polices which were predominated by rural landowners and agricultural producers (Lofchie, 1997). Rural development planners tended to view cities as parasitic or alien to rural interests, having a "rural bias," with "little or no interest in investigating how cities might be better brought into rural planning frameworks" (Douglass, 1998). Karshenas (1996) even dropped the use of the terms of urban bias and rural bias and switched to the use of a discourse about "industrial bias" or "agricultural bias."

In sum, in the developing world, many governments favor urban interests over rural development, and existing explanations focus on the political incentives of rulers to align themselves with powerful urban groups at the expense of rural residents (Pierskalla, 2011). For instance, Bates (1981) argued that the failure of Africa's agricultural policy was in part caused by adverse incentives for national political leaders. However, in reality, the urban–rural relationship was too complex to simply divide the national policy into "urban bias" or "rural bias," and it was difficult to measure the degree of bias. Rulers have to balance the threat of urban and rural residents by choosing a level of urban bias that minimizes the chance of insurgency (Pierskalla, 2011). As a consequence, policymakers have to periodically adjust their policy priority according to social, economic, and political situations. In China, for example, after more than two decades of urban priority development since the reform opening, the central government introduced the strategy of urban–rural coordinated development in 2003.

2.3 Urban–rural linkages and flows

The perceived link between city and countryside is evolving rapidly, shifting away from the assumptions of mainstream paradigms to new conceptual landscapes where rural–urban links are being redefined. Realizing the problems and potential social and economic risk, scholars have addressed the significance of urban–rural linkages since the 1980s. "Secondary cites," "The expanding cities," and "The networked model" are three representative theories of urban–rural linkage and flows. The urban–rural linkage has been further developed since the beginning of the 21st century.

2.3.1 Secondary cities

Rondinelli (1983) argued that the influence of urban bias theory and increasing poverty in rural areas played a determining role in a certain "neglect of cities" in development policy during the 1970s and 1980s. He proposed the theory of "secondary cities," which pointed out the failure of these strategies was mainly because of a lack of a system of cities which support economic activities and decentralized administrative functions necessary for successful development between urban and rural sectors. Meanwhile, rural–urban linkages were to be considered as promoting this widespread development. Only a decentralized investment in "strategically located settlements" could provide the necessary conditions to a "bottom-up" and autonomous development of rural communities. Unwin (1989) confirmed the contribution of Rondinelli's model in terms of considering the urban–rural interaction in that it relies heavily on "linkages both between rural areas and small cities, and on those between smaller and larger cities."

A critique from Stöhr (1981) pointed out that Rondinelli's strategies were basically top-down, and their goals of equity were difficult to achieve. However, "development from below" has been an alternative strategy of development, and it mainly harnesses local human and institutional resources, and initiatives in order to achieve the primary objective of satisfying the basic needs of that area. This strategy would be basic needs–oriented, regional resource–based, labor-intensive, small scale, and rural-centered. Meanwhile, it implies the use of an appropriate labor-intensive rather than a sophisticated capital-intensive technology. The main aim would not be development for development's sake, but development for the sake of people in general and of the poor in particular (Bhalla, 1982).

Unwin (1989) believed that many studies on "development" separating the urban and rural areas from one another resulted in a biased understanding of poor areas and the process of economic and social change. Therefore, he emphasized the interaction between urban and rural areas, suggesting that there were economic, social, political, and ideological linkages between urban and rural places on a broad theoretical level, and found that their physical expression in measurable flows of people, money, and budgetary allocation is associated with interactions between people, places, and objects. Although the scheme is unrealistic, flows are defined as physical and measurable, among them are "power" and "authority," while "ideas" find their place among the ideological flows. Moreover, Unwin set off a new trend of focusing on "urban–rural interaction" in the research of urban–rural development, and provided a general theoretical framework for future research (Adell, 1999).

Different from Unwin's sterile division in interactions, Douglass' model was more precise and suggested that research components can be divided into structures and flows. Rural structural change and development are linked to urban functions and roles through a set of flows between rural and urban areas, including patterns of flows, and their combined impact on fostering rural regional development (Douglass, 1998). Moreover, Douglass identified five types of flows: people, production, commodities, capital, and information, and each of these have multiple components and impacts. The five types of flows also feature different spatial linkages patterns and variable benefits to rural and urban areas.

2.3.2 The expanding city

Strong empirical evidence of metropolitan development patterns in market-oriented economies, in both industrialized and developing countries, illustrates there has been a world-wide phenomenon to move towards dispersal and growth on the peripheries or fringes of cities (Ingram, 1998). There is unanimous agreement that population growth in large cities usually promotes densification of less developed areas and expansion at the urban fringe (Ingram, 1998). Moreover, the growth of such agglomerations has been constantly enhanced by major improvements of technology in areas such as transportation and telecommunications (Jones & Visaria, 1997).

In Latin America, the poor have contributed dramatically to the creation of a lower class peri-urban fringe around large cities. This way of segregating lower income populations in the extensive and precarious peripheral areas has been prevalent where the "capitalistic sector" has allowed them to settle and become "home owners" (Ribeiro & Correa do Lago, 1995). Firman (1996) described that land was rapidly converted from agricultural to industrial and residential uses driven by investments in the extended metropolitan regions in Southeast Asia. An influx of population into these peripheral areas boosted land speculation and fostered prevalent and dynamic development activity, sometimes informal (Adell, 1999).

2.3.3 *The networked model*

Having analyzed the various development theories in the past which divided urban and rural areas, Douglass proposed the regional network development model from the perspective of urban–rural interaction. Douglass (1998) strongly argued a new paradigm for rural–urban linkages: the network model, which was based on a clustering of many settlements, each with its own specialization and localized hinterland relationships in contrast to making a single large city into an omnibus center for a vast region.

In a way, Douglass' networked model was similar to Rondinelli's secondary cities paradigm. Rondinelli considered that secondary cities should have an adequate size to perform decentralized activities, and they could be a part of a network of similar cities and of smaller ones. Compared with Rondinelli, Douglass' work was more realistic in his concept of the network and cluster ideas. He recognized the need to upgrade infrastructure both at the rural and urban level to achieve the necessary connectivity of the network (Adell, 1999). Upgrading local infrastructure services was essential for the quality of regional daily life and economic growth(Douglass, 1998).

Generally speaking, sustainable development of rural–urban interaction in the developing world has been identified as a key global challenge for the 21st century (Lynch, 2004). The rapid growth and mobility of urban population in the third-world has led to new changes in urban–rural linkage (World Bank, 2000). The urbanization process refers to a complex and dynamic rural–urban relationship which varies in different socio-economic contexts. In industrialized and developed Western countries, urbanization has followed a process of transition that assumes a classic model of rural–urban dichotomy (McGee, 2011). According to this concept, rural and urban areas are clearly divided spatially (Montgomery, 2003; Champion & Hugo, 2004; McGee, 2011). However, this conventional transition theory built on Western experiences is considered unable to reflect the urbanization process of contemporary developing countries which occurs in the unique background of time-space compression (McGee, 1991, 2011; Adell, 1999; Sui & Zeng, 2000).

In the present era, various kinds of flows are accelerated by telecommunication and globalization, of which flows between rural and urban areas are a part. Recently, based on Unwin (1989) and Douglass (1998) theory, many scholars have begun to reconsider flows and linkages between rural and urban areas. Satterthwaite and Tacoli (2003) paid particular attention to the ways in which modern economic, social, and cultural changes effect urban–rural interactions and built a model of "positive and negative" urban–rural interaction and regional development. This model emphasized the role of small and intermediate urban centers in rural and regional development and poverty reduction. Lynch (2004) proposed five aspects of flows "food, natural, people, ideas and finance" based on the interaction between urban and rural areas in developing countries. The core of the different performances of urban–rural interaction in different countries and regions was to reveal the impact of various "streams" on urban and rural areas in order to form policies and measures to alleviate poverty. Lynch also put forward the concept of "rural–urban dynamics" and suggested that the complexity of urban–rural linkages should be revealed from the perspective of "livelihood strategies" and "resource allocation." According to McGee (2011), rural–urban "interaction and linkage is a more accurate reflection of reality than the idea that rural and urban areas are undergoing somehow spatially separated transitions."

2.4 Globalization and extended metropolitan regions

Since the 1990s, as new forms of economic organization, technological change, and developing globalization, the metropolitan (Ingram, 1998) and the Extended Metropolitan Region (EMR) (McGee, 1997) emerged in the developing world, rural–urban linkages became much more multi-faceted, multi-layered, and spatially far-reaching than received spatial models.

2.4.1 Globalization and localization underlying the formation of Extended Metropolitan Regions (EMR)

Since the new century, progress in both theory and real-world contexts, in particular, space-time compression and globalization, has addressed the need of reassessing the changing nature of the rural–urban divide (Adell, 1999).McGee (1997) argued that with the globalization process, urbanization has been inevitably increased and global and sub-global systems of highly connected cities emerged. In the developing world, mega-urban regions appeared as major components of their urban systems. In the EMR, we find the trends of changing land use of the inner cores, residential outward movement, the creation and amelioration of transportation networks, and industrial decentralization into new industrial states have been popular.

From a broader regional perspective, urban linkages between these mega-urban regions have been enhanced as a result of growing economic interaction. A further consequence has been the emergence of sub-global regions: regional blocs of states within the global system. As an effect of globalization, population and economic activities are more concentrated in these EMRs which are regarded by both global and national investors as ideal areas to locate manufacturing of consumption goods, investing in building up the environment and as places to "create the landscape of global consumption" (McGee, 1997).

EMRs form in the process of urban expansion which are themselves structured by the forces of globalization and localization. The various combinations of the forces have led to different patterns of urban expansion and forms of EMRs. As highlighted by some research, influential factors of urban sprawl in the United States by Brueckner (2000) are quite different from what Wei and Ye (2014) summarized to cause China's urban sprawl. This indicates that the varied urban sprawl patterns are not attributed to the process *per se*, but are due to the distinct institutional framework in a given geographical context (Hall & Jones, 1999).

In Western developed countries, such as the United States, increasing suburbanization due to rising income, lower cost of transportation, and the American lifestyle of addicted driving (Brueckner, 2000) has caused their urban sprawl patterns. This has resulted in scattered commercial strip development in suburbs and the decline of the traditional urban center (Black, 1996; Geddes, 1997). However, in the context of developing Asian countries which have been experiencing dramatic economic transformation and institutional change during the constantly strengthening globalization process, dramatic and uncontrolled urban land expansion prevailed from the urban center to suburban districts with unprecedented scale and speed (Fung, 1981; Tian, 2014). The obvious manifestations of this expansion dot the Desakota landscape (McGee, 2011) widely identified in Southeast Asia and manufacturing driven peri-urbanization in the coastal areas of China (Webster & Lai,2003).

In the context of China, its internal responses to global change can be summarized as marketization with land as a crucial domain and power decentralization among different hierarchies of governments and government agents. Chinese cities used to be compact and land was a means of production allocated by the central government according to political ideology through a project-specific land allocation system (Yeh & Wu, 1996). Under this central-planned system, urban sprawl was rarely found in the country. After 1978, triggered by the profound economic reform incorporated with institutional changes from the central state to local municipalities, urban sprawl occurred in a tremendous urban land expansion process across Chinese cities. This sprawl occurred within a land-dependent growth model beginning in the Pearl River Delta (PRD) region during the 1980s and spread nationally in a horizontal perspective (Wu & He, 2005). To be specific, accompanied with the land reform and power decentralization process, the restructuring of urban land use was widely implemented across different regions and cities, which was characterized by redevelopment of urban cores, relocation of industries from the urban cores to suburban areas, and the implementation of large-scale housing construction in the urban fringe (Yeh & Wu, 1996; Seto & Fragkias, 2005).

In general, it has been commonly accepted that China's urban expansion is deep-rooted in its political-economic transformation. The Chinese version of urban sprawl is a deliberate, ambitious project led by proactive municipalities to acquire land, build new development districts (*Kaifaqu*) such as industrial parks and high-tech economic development zones at the urban periphery, as well as to build large-scale residential communities (Wu & He, 2005). Specifically, as representatives of the state, local governments not only own the urban land within their jurisdiction but are also decision-makers of city master plan, land-use plan, as well as relevant policies and regulations (Lin et al., 2010; Zhao et al., 2010).

Since GDP growth is a crucial indicator to evaluate performances of local officials and helps in deciding whether they can be promoted to higher political positions, local governments are inclined towards local development and have strong incentives to pursue local growth through the tool of land-use policy (Zhu, 2004). On the one hand, local governments are motivated to supply cheap land for industrial uses and attract foreign investments to stimulate economic growth, utilizing land transactions with commercial and residential developers to generate land sale revenues (Lin et al., 2010). On the other hand, through administrative adjustments and the construction of special development zones, lands from surrounding counties and suburbs are under the control of higher level urban governments and serve their plans for development (Yang et al., 2009).

2.4.2 Formation of China's typical EMRs: PRD, YRD, and BTH

Despite the aforementioned general explanation of China's urban expansion, the patterns of urban expansion and the forms of metropolitan areas within China vary in different socio-economic and political circumstances and are locale-specific. This section is devoted to discussing three most dynamic EMRs in China: Pearl River Delta (PRD), Yangtze River Delta (YRD), and Beijing, Tianjin, and Hebei region (BTH)(Figure 2.2). PRD, located in the middle and south part of Guangdong Province, covers 14 cities, including Guangzhou, Shenzhen, Zhuhai, Foshan, Dongguan, Zhangshan, Jiangmen, Zhaoqing, Huizhou, QingYuan, Yunfu, Yangjiang, HeYuan, and Shanwei city. The land area of PRD takes a share of 0.57% of the national territory. In 2016, its population accounted for 4.23% of national total population, and its GDP reached 7,311.877 billion Yuan, accounting for 9.2% of national total GDP. Having Shanghai as its center and Nanjing, Hangzhou, and Hefei as its sub-centers, YRD includes 26 cities such as Shanghai, Suzhou, and Changzhou, and its proportion of land area, population, and GDP in the country was 2.2%, 19.8%, and 11.0% in 2016, respectively. The BTH region is located in the heartland of the Bohai Bay Economic Zone in Northeast China, including two municipalities directly under the Central Government: Beijing, Tianjin, and 11 prefecture-level cities such as Langfang and Shijiazhuang. In 2016, the proportion of land area, population, and GDP of BTH in the country was 2.2%, 10%, and 7.97%, respectively.

As mentioned by Zhao et al. (2010) and Kuang et al. (2016), the degree to which a particular location has been influenced by institutional changes in terms of national policies and market forces greatly determines spatial growth patterns and urban forms. For example, literature using remote-sensing data to measure China's urban expansion identifies three distinct stages (Xiao et al., 2014): between 1981 and 1998, the national urban expansion rate remained quite low. From 1998 to 2008, the urban expansion rate increased dramatically. The final stage of urban expansion found in such literature peaked after 2008.

As for the three EMRs, the PRD region has experienced the highest rate of expansion and this region also began expanding the earliest (1985). The east coastal region (YRD) ranks second in overall expansion and has

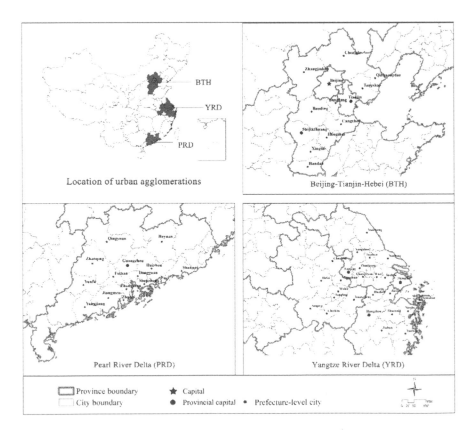

Figure 2.2 Location of three EMRs and their composition
Source: drawn by authors

expanded most rapidly of any region since 1998. The expansion process of BTH lagged far behind that of the PRD and YRD regions, with one notable expansion from 1998 to 2000. Other regions' expansion rates remained at a low level, with an apparent upward trend beginning in the northeast, southwest, and northwest regions in 2008. These stages generally coincided with the spatial-temporal implementation of national policies: implementation of reforms and openness in the PRD region in the early 1980s, development of Pudong, Shanghai in the early1990s, and development of the Tianjin seafront district in the early 2000s. The following sections examine the spatial-temporal patterns and driving forces of expansion of built-up area in three typical metropolitan regions.

2.4.3 PRD region: leading urban expansion with obvious sprawling patterns and a fragmented land development process

As the earliest area in which experimental economic reform was implemented, the PRD region's pioneering urban growth and land expansion in the 1980s was to be expected. Correlating well with the advantages of its coastal location and favorable policies, the dramatic urban expansion, as well as the urban sprawl patterns in the PRD, has been identified in a considerable amount of studies. The quantitative study of spatial-temporal patterns on urban land-use change of four cities in the PRD conducted by Seto and Fragkias (2005) revealed several significant features of the urban sprawl process in this region. They applied remote sensing data (Landsat TM images) and landscape metrics to analyze urban growth from 1988 to 1999 across three buffer zones (0–3 km, 3–10 km, and 10–20 km) in Guangzhou, Shenzhen, Dongguan, and Zhongshan. The development pattern of multi-nuclei was confirmed over each of the studied cities, followed by the edge-expansion mode to connect surrounding landscapes.

From 1988 to 1991, there was an obvious trend of new sub-center development in terms of "the leapfrogged model" in Shenzhen. From 1991 to 1996, edge-expansion was a dominant urban expansion pattern which shaped a coalescence process of dispersed land patches. Notably, the self-contained land redevelopment of rural villages and towns, as well as governments' construction of industrial zones with a huge inflow of foreign investment under loose policy and political controls, enabled the urban form of the PRD region to change quickly over a short time. Much of this change occurred in the urban patterns in Shenzhen and Dongguan between 1992 and 1996. The results suggest the region benefited from minimal zoning restrictions. As demonstrated, even a rural area with basic infrastructure can be rapidly expanded, creating fragmented spatial patterns within the strong market forces if there are minimal policy and political restrictions. The authors argued that the cities in this region generally followed similar spatial expansion patterns, which were fragmented and greatly driven by foreign investment.

More specifically, Chen et al. (2014) compared two adjacent cities in this region, Shenzhen and Dongguan. This study looked at the cities from 1990 to 2008 in terms of their urban expansion modes, shaped spatial patterns and driving forces, and employed satellite images to analyze spatial-temporal changing patterns, as well as government annual statistical reports on population, industry, transportation, and policy to explore their impact on urban expansion. Both cities were shown to have a rapid increase in urban expansion and urban expansion intensity from 1990 to 1999, and a decreasing trend from 1999 to 2008. The process of infilling and edge-expansion of construction land was remarkably strong between 2005 and 2008, suggesting a gradual compact urban development form in the late 2000s. While analyzing the driving forces of urban expansion, the authors argued that Shenzhen's upgrading of industrial structure was coordinated by the government at the district or municipal level instead of Dongguan's town-level government, and Shenzhen's urban development was more sustainable than that of Dongguan.

The spatial-temporal patterns of urban expansion have changed dramatically in the PRD region since China's economic reform, with an emerging multi-nuclei urban development form identified across major cities. In particular, differences in the trajectory and intensity of urban expansion exist among its different cities. This region experienced fragmented land expansion patterns in sub-centers and rural areas before 1990 and a relatively compact expansion process of "infilling" and "edge-expansion" after 2000 in-between urban centers, sub-centers, and development zones. From 1990 to 2000, the two modes coexisted, as seen in the redevelopment process in urban centers and in new land development around sub-centers which were driven by a continual inflow of foreign investment. Rapid urban expansion and fragmented urban development resulted from a huge inflow of foreign investment brought about in part by national policy in post-reform China. This correlated well with good location and minimal land-use restrictions by local governments in this region.

2.4.4 YRD region: compact and dramatic urban land development under strong government coordination

Referred to as the other dominant economic zone in China, the YRD region has experienced a dramatic and intense urban expansion process since the 2000s, more than any other region since the central state adjusted preferential policies to support development in this area. The form of expansion patterns and urban development in these cities, however, differed greatly from those of the PRD cities. Contrary to a fragmented urban development form and the fast-growing sub-centers in PRD cities, Chen et al. (2015) identified circular expansion patterns in a compact form based on a study of 51 YRD cities from 1979 to 2007. The authors pointed out that cities in this region displayed a compact centralized form with rapid expansion around a relatively small mature central urban area. Additionally, the period of more dramatic urban expansion for all cities occurred from 2001 to 2007. With their mature urban centers and large-scale expansion, Shanghai, Nanjing, Wuxi, Suzhou, and Hangzhou played a dominant role in both the intensity and trajectory of urban expansion in the YRD region. Some multi-core spatial patterns, as well as a linear spatial shape of urban expansion, have been found in cities such as Taizhou, Ningbo, and Huzhou of Zhejiang province and Taizhou and Zhenjiang of Jiangsu province, indicating that sprawling urban expansion patterns emerged in these medium-sized cities. The authors concluded that rapid urban development and multi-center urban form led to the sprawling land-use pattern.

The picture in the YRD was depicted by Gao et al. (2015), who employed spatial regression models, as well as geographically and temporally weighted models to analyze demographic and socioeconomic data for YRD cities from statistical data books between 2000 and 2010. They discovered that although industrial and residential land expanded rapidly in the YRD cities, the developed areas in Zhejiang and Jiangsu provinces began to coordinate their urban land-use structure in a compact way. However, simple, low-density land-use patterns were still dominant in less developed areas of the two provinces. Findings revealed that county-level cities continued to focus on industrialization while prefectural cities transformed their structure of industry from production to service-based. This suggests that, to a great extent, either the variation of administrative levels of government in coordination with urban land uses, or the profit-seeking in the industrial selection by the government, determines the spatial patterns of urban growth.

Indeed, Tian et al. (2011) have already compared the urban expansion trajectory of five dominant urban areas, including Shanghai, Nanjing, Suzhou, Wuxi, and Changzhou in this region through an analysis of landscape metrics and concentric buffer zones across these cities. All five urban areas experienced dramatic urban land expansion from 1990 to 1995 and 2000 to 2005, with an obviously slow expansion process between 1995 and 2000. In particular, Shanghai, Suzhou, and Changzhou shared a similar urban expansion trajectory, reflected as a coalescence process. The main growth types were "infilling" and "edge-expansion" from 1990 to 1995, followed by a diffusion process from 1995 to 2000 and a coalescence process, including edge-expansion and infilling, between 2000 and 2005. However, the process of coalescence throughout all 15 years in Nanjing and Changzhou suggested a continuously compact expansion within these two cities.

Compared with spatial change patterns and urban expansion modes in the PRD region, major cities in the YRD area, including Shanghai, Nanjing, Suzhou, Wuxi, and Changzhou, were discovered to have a centralized compact form of urban expansion. This was reflected in the dominant urban expansion modes of infilling and edge-expansion and the circular land expansion process around a mature urban center. More importantly, the role of local government, especially in the coordination of urban land-use structure, was revealed to ultimately drive the compact patterns of urban expansion in this region.

2.4.5 BTH region: Beijing–Tianjin-centered land expansion with a dual-form urban development pattern

When considering a region which includes two direct-controlled municipalities, with one of those municipalities being the Chinese capital, this region is determined to be bathed in a dense political nature. The BTH region, therefore, never falls behind in terms of economic or population growth, or in the process of urban land expansion. The entire region experienced rapid urban expansion from 1990 to 1995 and a slowdown trend from 1995 to 2000 according to Tan et al. (2005) who conducted research on the area between 1990 and 2000. They argued that the land expansion rate was much higher than the rate of population growth, and that the slowdown trend of urban land expansion was attributed to stricter land-use regulations and policies during this time period, suggesting that urban sprawl patterns and land-use policies had a big effect on driving the urban expansion process. It is worthwhile to note that this impact from policies is much larger in small cities than that in large cities, indicating such policies may not be fully implemented across an entire megacity area like in Beijing and Tianjin.

Incorporated with GIS techniques, Landsat images, and landscape ecological analysis, Wu et al. (2013) implemented a comparative study of spatial-temporal patterns during the process of urban expansion (1980–2010) in three major cities of the BTH region, Beijing, Tianjin, and Shijiazhuang. Regarding urban development patterns, they defined Beijing's urban spatial form as a mononuclear polygon since the city's urban land expanded from its central city in all directions. A double-center spatial pattern was identified in Tianjin, reflected in the rapid development in the southeast coast of Tianjin since 1995 which has become a gradually mature sub-center connecting with the original urban core of the city. Shijiazhuang was represented by a number of small growth points located away from the urban center, shaping a point urban expansion pattern.

The transformation process from outlying growth to infilling identified in all three cities confirmed they were becoming more compact. Specifically, outlying growth mainly occurred from 1980 to 1990 in Beijing, corresponding well with the satellite towns distributed in the rural area away from the urban core. Beginning in 1990, infilling became dominant in Beijing's urban expansion along its ring roads, associated with edge-expansion in suburban areas around the urban core. For Tianjin, the shift from outlying growth to infilling began in 1995, when construction began on the Binhai New Area. In Shijiazhuang, around 2000, the focus of urban expansion turned the scattered points to the aggregated growth in the urban core. The authors highlighted the role of urban planning and policies in guiding and controlling the process of urban expansion among the three cities. In addition, they pointed out regional disparity, indicating that Beijing's strong institutional background imposes a negative "shadow effect" on other cities including Shijiazhuang and even the early period of expansion in Tianjin. "The poverty belt around Beijing-Tianjin" described the double-central urbanization pattern in the BTH region, where Beijing and Tianjin play such a large role that other cities in this region become the hinterland for these two cities.

2.5 Peri-urbanization: concepts, models, and spatial patterns

2.5.1 Peri-urbanization: concepts and definitions

Since the 1940s, urban studies scholars have begun to focus their research on the urbanization processes that were shaping the peri-urban interface. During the 1940s and 1950s, the term "urban fringe" was widely adopted in academic literature, meaning the area where suburban growth took place and where urban and rural uses of land were mixed, forming a transition zone between city and countryside (Thomas, 1974).

During the 1980s and 1990s, some new concepts and terms were created to help better describe contemporary forms of urbanization: megalopolis, techno-burbs, edge city, peperoni-pizza cities, superburbia, a city of realms, perimeter cities, outer cities, heteropolis, and peri-metropolitan bow waves (Oatley, 1997). Meanwhile, the shift to conceptualizing urban–rural interaction as a part of a new landscape echoes Europe, as is found in the terms of peri-urbanization and rurbanization in France (Adell & Capodano, 2001), and the Zwischenstadt in Germany (city-in-between: Sieverts, 1999). The emergence of these new landscapes has challenged professional perceptions and knowledge across Europe, and the new concept has raised questions about the necessary adaptation of the urban and regional planning set of conceptual and practical tools to cope with the new territorial reality across the continent (Adell, 1999).

According to concepts, peri-urban areas are basically considered a transitional zone between urban and rural areas, which are always difficult to define and filled with problems inherent to both rural and urban worlds. As a transitional area, this affects the very definition of such an entity as a peri-urban (or rural–urban) interface, which has shifted from a spatial definition to a more functional focus on diverse flows between rural and urban sectors. The definitions of these two visions have been gradually replaced by other new concepts which are different from the traditional discourses on "peri-urban" areas.

The first example is the rural–urban fringe proposed by Carter (1981): the space into which the town extends as the process of dispersion operates, an area with distinctive characteristics which is only partly assimilated into the growing urban complex, but still partly rural, and where many of the residents live in the country. However, the residents are not rural any longer, either socially or economically. The second case comes from Webber (1964), who proposed the concept of the urban realm: an urban realm is neither urban settlement nor territory, but heterogeneous groups of people communicating with each other through space. This was based on the fact that the basic functions of the city are transferred from the central city to the suburbs and then to the further, largely realized area of the "urban field," where the mobile middle classes have built a highly dispersed pattern of activities (Adell, 1999).

More recently, with the global transition, a new lexicon has emerged that makes reference to either a new type of urban development which is characterized by employment and other activity concentrating at some distance from old urban centers, or stressing the processes underpinning the new developments (in particular, the impact of flexibility in production systems and technology) (Oatley, 1997). It was argued that the central transition in urban form involves a shift from the compact/suburban dualistic pattern to the spread of metropolitan areas and from mono-centric cities to multi-centered urban areas (Garreau, 1992; Hall, 1996).

Following that, terminological and conceptual research took place. Kurtz and Eicher (1958) regarded urban fringe and suburbs as different concepts, and Wissink (1962) further classified the urban forms into pseudo-suburbs, satellites, and pseudo-satellites, and inner and outer urban fringe areas. Then, Pryor (1968) distinguished between peri-urban areas and the urban periphery. Another definition of "peri-urban" areas moving away from a physical dimension shows it being the result of particular social processes, mainly the migration of mobile middle class families oriented to the city and dominated by urban lifestyles (Pahl, 1965).

The commonly accepted factor that makes the rural–urban interaction more prominent in developing Asian countries is the high population density. In these countries, urbanization takes place in the form of urban activities penetrating into the already densely populated rural areas (Sui & Zeng, 2000; Zhu, 2013). In the contexts of both time-space compression and high population density, rapid urbanization in these countries has resulted in the prevalence of peri-urban areas. According to Rakodi (1998), a

> peri-urban area is a dynamic zone both spatially and structurally. Spatially, it is the transition zone between fully urbanized land in cities and areas in predominantly agricultural use. It is characterized by mixed land uses. It is also a zone of rapid economic and social structural change.

Peri-urbanization is a term coined to describe the process of forming peri-urban areas. It has been claimed by Webster and Muller (2002) that peri-urbanization has become a dominant form of urbanization in developing Asian countries and will continue to have great impact, accounting for up to an 80% increase in urban areas over the next few decades.

2.5.2 The Desakota model

In the late 20th century, areas where agricultural and non-agricultural activities are intermixed appeared on the fringe and traffic corridors between the core cities in Asia, and they formed complex and compound regional systems consisting of central cities, fringe areas of those cities, exurbs, satellite towns, and extensive intervening areas of dense population and intensive traditional agricultural land (Ginsburg et al., 1991). In 1997, McGee proposed a territorial model named Desakota to describe the intense mixture of non-agricultural and agricultural activities in these regions. According to McGee's model, five main regions were identified (Figure 2.3):

- Central cities (in the Asian context, this is generally a very large city);
- Peri-urban regions: those areas surrounding cities within a daily commuting distance from the city center;
- Desakota regions: they usually lie along the corridors connecting a large city center to smaller town centers;
- Densely populated rural regions;
- Sparsely populated frontier regions.

2.5.3 Peri-urbanization in China: governance and spatial patterns

This section offers a comprehensive summary of models of China's peri-urbanization in terms of urban expansion patterns and related governance structures which have been researched through existing literature with case studies. Generally speaking, different urban expansion processes are shaped

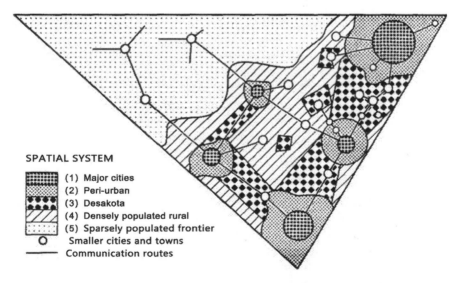

SPATIAL SYSTEM

- (1) Major cities
- (2) Peri-urban
- (3) Desakota
- (4) Densely populated rural
- (5) Sparsely populated frontier
- ○ Smaller cities and towns
- —— Communication routes

Figure 2.3 An illustration of Desakota model
Source: public domain

by varied circumstances of globalization and local political-economic and social environments. Three modes of peri-urbanization are identified: (1) the PRD region, which has witnessed fragmented peri-urbanization mainly characterized by village-led rural industrialization, along with top-down land development by urban governments; (2) the YRD region, of which urban expansion is mainly coordinated by governments at the town level or above, and peri-urbanization is relatively compact and shaped by government-initiated development zones, industrial parks, and new towns; (3) The BTH region where peri-urbanization is mainly shaped by urban development is coordinated by the governments at the district level or above.

(1)   PRD region: fragmented peri-urbanization led by rural industrialization

The key importance of pioneering institutional reform in the PRD area, indeed, was already emphasized by Yeh and Wu (1996) who examined Guangzhou's spatial pattern changes of land development and driving forces before the land reform (1979–1987) and after it (1987–1992). Their research was based on a combination of remote sensing data, GIS, and a logistic regression model. Before the land reform, industrial land development was the dominant type of land converted from agricultural land. However, after the reform, commercial and residential land uses became more significant, leading a redevelopment process to emerge in the urban center. Meanwhile, new sub-centers appeared in the peripheral areas, such as economic and technological development zones, accompanied with construction of large-scale housing projects. The driving forces underlying urban land development from 1979 to 1987 were cheaper land in suburban areas relative to city centers, close proximity to highways, and close proximity to industrial parks. However, between 1987 and 1992, the proximity to the urban center, sub-center (development zones), and city streets were more significant in shaping land development patterns. According to their research work, the changes of urban land-use structure and the prosperous development of sub-centers in Guangzhou were significantly highlighted after the institutional reforms.

Regarding another piece of research work, Yu and Ng (2007)utilized two urban–rural transects with landscape metrics to examine the spatial configuration from the urban center to rural areas from 1988 to 2002. They identified fragmented and leapfrogged land expansion patterns at the urban fringe of the city or new urbanizing areas in Guangzhou. Designated as a pioneering area to experience national reforms, a test area, cities in the PRD region practiced land development under a special economic regime (Lin et al., 2010), strong market forces, and a large inflow of foreign direct investments (Zhu & Guo, 2014).

Accompanied with strong market forces, however, was a massive land development process based on the dual land tenure system between urban land and rural land brought about by the institutional reform (Zhu, 2005). On the one hand, municipal governments were motivated to pursue local growth through land development, significantly driving urban sprawl patterns extending to rural areas. On the other hand, from a bottom-up perspective, rural collectives and township governments occupied a considerable area of land based on the historical development of township-village enterprises, enabling them to self-develop land under the loose control of high-level government under PRD's special institutional regime (Lin & Ho, 2005; Lin et al., 2010). As a consequence, informal land development coexisted with government-led urban sprawl patterns, creating unique spatial patterns such as peri-urban landscape and the phenomenon of urban villages.

An urban expansion mode characterized as a combination of a relatively weak central city and greatly developed but fragmented townships has been explored in the cities of the PRD region (Yu & Ng, 2007; Tian & Liang, 2013; Zhu, 2017). Two modes of land development in this region, those being top-down and bottom-up land development, respectively, are regarded as the deep-rooted factors for expansion by a number of studies, such as the analysis of urban villages in Guangzhou conducted by Zhou (2014), the fragmented peri-urbanization, and informal land institutions in Nanhai explored by Zhu and Guo (2014), and the bottom-up rural industrialization and fragmented governance in Shunde analyzed by Tian (2015).

(2) YRD region: compact peri-urbanization formed under the coordination of town-level and higher governments.

Since the 1990s, national preferential policies have been adopted in the YRD area. These are associated with its advantageous location and the historical strong base of township-village enterprises, which are similar to the political and economic background of the PRD region. Yin et al. (2011) studied urban expansion and land-use/cover changes in Shanghai from 1979 to 2009 and confirmed a continuous and accelerated urban expansion process across three time periods (minor expansion between 1979 and 1990, rapid expansion from 1990 to 2000, dramatic expansion between 2000 and 2009). This generally matched the conclusion drawn by Shi et al. (2009) that urban expansion was dominant in Shanghai from 1992 to 2006. Specifically, before 1990, urban expansion was concentrated on a north–south axis. However, from 1990 to 2009, expansion of construction land occurred in all directions along main transportation routes between the city center and surrounding towns. These two studies highlighted the key role of economic restructuring and government control, which is consistent with the analysis of urban expansion patterns in peri-urban areas of Shanghai conducted by Tian et al. (2017).

This most recent study identified the planned growth of urban sprawl patterns in terms of the construction and development of industrial parks and new towns located in peri-urban areas. This confirmed the significant role of government's top-down plans in directing the urban expansion process in Shanghai from 1990 to 2009. Indeed, government-led urban sprawl patterns have also been discovered in other major cities located in the YRD region, such as the tremendous urban expansion and sprawl process highlighted in Hangzhou (Yue et al., 2010; Wu & Zhang, 2012). Identified as industrial-oriented urban sprawl, Hangzhou's urban expansion and its polycentric patterns were largely attributed to the evolution of land institutions and development controls by governments, as well as the gradually strong market forces in the coastal area (Qian, 2008).

Urban expansion patterns discovered in the YRD region, a relatively compact urban sprawl mode, displayed as a mature city center with circular edge-expansion patterns along transportation corridors to link satellite towns (Shanghai), as well as government-led urban sprawl in terms of urban jurisdiction adjustments (Hangzhou), have been widely explored in existing literature. While confronting similar political and economic conditions as those in the PRD region, specifically the coastal area with prior national policies and strong market forces, major cities in the YRD region demonstrated a totally different urban sprawl mode. This indicates the key importance of local government policies and administrative structure in determining distinct urban sprawl patterns. The strong market forces and preferential development policies brought a tremendous amount of foreign investments and a rapid process of industrialization to the YRD region. However, the strict coordination of urban land-use structure and economic restructuring in major cities constrained urban sprawl patterns to a compact form in the dramatic process of urban expansion.

To be specific, compared with the strong awareness of property rights at the village level in the PRD region, the practice of land rights and land development in the YRD region has been concentrated at the town level or above (Zhu, 2017). Through a coordinated plan and cooperation between government and rural communities, urban land expansion and construction activities were generally under a unified planning system, reflected in the establishment of satellite towns and industrial parks.

An urban form combining a mature center with multiple sub-centers and strong towns has gradually emerged under the control and coordination of the municipality. In this sense, a peri-urbanized area of dominant development zones and new towns has evolved in Shanghai, creating a sharp contrast with the complex mix of rural industries and informal residential land use in peri-urbanized area. Additionally, the comparative analysis of Kunshan and Nanhai proposed by Zhu and Guo (2014) illustrated different urban land development modes between the YRD and PRD regions. However, the adjustment of administrative boundaries and development policies by the municipal government of Hangzhou, as well as county-governments' seeking of industrialization and land development in suburban areas (Chen et al., 2014), have reshaped the urban structure and created distinct urban sprawl patterns in these areas.

(3)   BTH region: peri-urbanization dominated by top-down land development

The previous section highlighted the formation of three outstanding examples of urban areas within China with high rates of expansion. We looked at the massive and disordered land expansion process under the fragmented governance of cities located in the PRD region, the relatively compact urban land development form with strong local government control and coordination of multidimensional urban structure across YRD cities, and the dual-form urban land expansion patterns with dense political characteristics in the BTH area.

Regarding the urban sprawl process in the BTH area, the political nature of the BTH region plays a crucial role in determining the dominant intervention of government over market forces. The latter has even evolved as a tool by governments to control land development (Zhu & Hu, 2009; Zhao et al., 2010). Du et al. (2014) analyzed the land market combining the ambitious urban land reform and rural–urban land conversion, and stated that the gradually emerging land market had a large impact upon land-use conversion in Beijing. In terms of spatial land-use changes, since 1992, inner suburban districts experienced a rapid loss of cultivated land. From 1996 to 2001, land development of suburban areas was dominant. Since 2002, outer suburban land development has experienced a rapid and dramatic increase in Beijing due to the spill-over effects of central city.

The conflict between municipal government's compact development strategy in terms of formalizing the land development market (Zhao et al., 2010) and the competition for land rent by the local town and township governments (Zhu & Hu, 2009) led to a compact centralized expansion pattern in the urban core and dispersed sprawling patterns in outer suburban areas of Beijing (Zhao et al., 2010). With regard to the more specific research work from Kuang (2012), it revealed that the function of the urban core area was transformed from economic and industrial use (before the 1990s) to culture, technology, and residence in Beijing. Meanwhile, suburban districts bore the responsibility of relocating industrial lands, the boom of residential construction, as well as the creation of green space. As a result, the central city of Beijing demonstrated a compact land development mode and the municipal government directed an urban containment strategy at the same time. However, as the rural–urban fringe of Beijing experienced opportunities for development brought about by the restructuring of urban land-use policy implemented by the municipality, a sharp conflict arose between the local growth objectives of suburban town/township governments and the compact development strategies proposed by the municipal government (Zhao et al., 2010). The dual-form urban sprawl mode, specified as the gradual compact central city with multiple functions, and the dispersed sprawling patterns in suburban districts driven by lower level governments' land development motivations have gradually emerged across Beijing's central city and suburban districts.

References

Adell, G. (1999).Theories and Models of the Peri-urban Interface: A Changing Conceptual Landscape. Strategic Environmental Planning and Management for the Peri-Urban Interface Research Project, University College,London.

Adell, G., & Capodano, X. (2001).Dire Les Nouveaux Territoires: du Stigmate de La Banlieue ÀL'ubiquitÉ du Paysage. *CARACTERITZACIÓ DE LA LITOSFERA I ELS SEUS PROCESSOS EN LA ZONA DE INTERACCIÓ DE LES PLAQUES TECTÒNIQUES D'IBÈRICA I ÀFRICA*– London: ResearchGate.

Bates, R. H. (1981).*Markets and States in Tropical Africa*. Berkeley: University of California Press.

Bhalla, G. S. (1982). Book Review: Development from Above or Below? The Dialectics of Regional Planning in Developing Countries. *Urban Studies*, 19:430–432.

Black, J. T. (1996).The Economics of Sprawl. *Urban Land*, 55(3):6–52

Brueckner, J. K. (2000).Urban Sprawl: Diagnosis and Remedies. *International Regional Science Review*, 23(2):160–171.

Carter, H. (1981).*The Study of Urban Geography*. Victoria, Australia: Edward Arnold.

Champion, T.,&Hugo, G. (2004).*New Forms of Urbanization: Beyond the Urban-rural Dichotomy*. Aldershot, UK: Ashgate.

Chen, J. F., Chang, K.-T., Karacsonyi, D., & Zhang, X. (2014).Comparing Urban Land Expansion and Its Driving Factors in Shenzhen and Dongguan, China. *Habitat International*, 43:61–71.

Dobrescua, E. M., & Dobre, E. M. (2014).Theories Regarding the Role of the Growth Poles in the Economic Integration. *Procedia Economics and Finance*, 8:262–267.

Douglass, M. (1998).A Regional Network Strategy for Reciprocal Rural-Urban Linkages: An Agenda for Policy Research with Reference to Indonesia. *Third World Planning Review*, 20(1):1–33.

Du, J. F., Thill, J.-C., Peiser, R. B., &Feng, C. (2014). Urban Land Market and Land-Use Changes in Post-Reform China: A Case Study of Beijing. *Landscape and Urban Planning*, 124:118–128.

Fei, J. C. H., & Ranis, G. (1997). *Growth and Development from an Evolutionary Perspective*. Oxford: Basil Blackwell.

Firman, T. (1996).Urban Development in Bandung Metropolitan Region: A Transformation to a Desa-Kota Region. *Third World Planning Review*, 18 (1):1–22.

Fung, K. I.(1981). Urban Sprawl in China: Some Causative Factors. In Ma, L. J. C., &Hanten, E. W. (Eds.) *Urban Development in Modern China*. Boulder: Westview Press, 194–221.

Gao, J. L., Wei, Y. D., Chen, W., & Yenneti, K. (2015). Urban Land Expansion and Structural Change in the Yangtze River Delta, China. *Sustainability*, 7(8):10281–10307.

Garreau, J. (1992).*Edge City: Life on the New Frontier*. New York: Anchor Books.

Geddes, R. (1997).Metropolis Unbound: The Sprawling American City and the Search for Alternatives. *American Prospect*, 8(35):40–46.

Ginsburg, N. S., Koppel, B., & Mcgee, T. G. (1991).*The Extended Metropolis: Settlement Transition in Asia*. Honolulu: University of Hawaii Press.

Hall, P. (1996).Revisiting the Nonplace Urban Realm: Have We Come Full Circle? *International Planning Studies*, 1(1):7–15.

Hall, R. E., & Jones, C. I. (1999).Why do some countries produce so much more output per worker than others?*The Quarterly Journal of Economics*, 114(1):83–116.

Ingram, G. K. (1998).Patterns of Metropolitan Development: What Have We Learned?*Urban Studies*, 35(7):1019–1035.

Jones, G. W., & Visaria, P. M. (1997).*Urbanization in Large Developing Countries: China, Indonesia, Brazil, and India*. Oxford: Clarendon Press.

Karshenas, M. (1996). Dynamic Economies and the Critique of Urban Bias. *Journal of Peasant Studies*, 24(1–2):60–102.

Kuang, W. H. (2012). Spatio-Temporal Patterns of Intra-Urban Land Use Change in Beijing, China between 1984 and 2008. *Chinese Geographical Science*, 22 (2):210–220.

Kuang, W. H., Liu, J., Dong, J., Chi, W., & Zhang, C. (2016). The Rapid and Massive Urban and Industrial Land Expansions in China between 1990 and 2010: A CLUD-Based Analysis of their Trajectories, Patterns, and Drivers. *Landscape and Urban Planning*, 145:21–33.

Kurtz, R. A.& Eicher, J. B. (1958).Fringe and Suburb: A Confusion of Concepts. *Social Forces*, 37(1):32–37.

Lewis, W. A. (1954). Economic Development with Unlimited Supply of Labor. *The Manchester School*, 22(2):139–191.

Lin, G. C. S., & Ho, S. P. S. (2005). The State, Land System, and Land Development Processes in Contemporary China. *Annals of the Association of American Geographers*, 95(2):411–436.

Lin, T., Guo, X., Zhao, Y., Pan, L., & Xiao, L. (2010). A Study of Residents' Environmental Awareness among Communities in a Peri-Urban Area of Xiamen. *International Journal of Sustainable Development and World Ecology*, 17(4):285–291.

Lipton, M. (1977).*Why Poor People Stay Poor: Urban Bias in World Development*. Cambridge: Harvard University Press.

Lofchie, M. F. (1997). The Rise and Demise of Urban-Biased Development Policies in Africa. *Cities in the Developing World Issues Theory & Policy*. 1997(3):23–39.

Lynch, K. (2004). *Rural-Urban Interaction in the Developing World*. New York: Routledge Press.

McGee, T. G. (1991). The Emergence of Desakota Regions in Asia: Expanding a Hypothesis. In Ginsburg, N., Koppel, B., & McGee, T. G. (Eds.) *The Extended Metropolis: Settlement Transition in Asia*. Honolulu: University of Hawaii Press, 3–25

McGee, T. G. (1997). Globalisation, Urbanisation and the Emergence of Sub-Global Regions. In Watters, R. F., & McGee, T. (Eds.) *Asia Pacific: New Geographies of the Pacific Rim*, London: Hurst and Company.

McGee, T. G. (2011). Managing the Rural-Urban Transformation in East Asia in the 21st Century. International Symposium on Urban/Regional Planning in Asia. *Journal of Urban & Regional Planning*, 3(1):155–167.

Montgomery, M. R. (2003).*Cities Transformed: Demographic Change and Its Implications in the Developing World*. Washington, DC: The National Academies Press.

Nolan, P.,&White, G. (1984). Urban Bias, Rural Bias or State Bias?Urban-Rural Relations in Post-Revolutionary China. *Journal of Development Studies*, 20(3):52–81

Oatley, N. (1997). *Lexicons of Suburban and Ex-urban Development*.Research Project, City Words Programme, Paris: UNESCO-CNRS.

Pahl, R. E. (1965).*The Metropolitan Fringe in Hertfordshire*. London: The London School of Economics and Political Science.

Perroux, F. (1950). Economic Space: Theory and Applications. *Quarterly Journal of Economics*, 64:89–104.

Pierskalla, J. H. (2011).A Theory of Urban and Rural Bias: A Dual Dilemma of Political Survival (Doctoral dissertation).

Pryor, R. J. (1968). Defining the Rural–Urban Fringe. *Social Forces*, 47(2):202.

Qian, Z. (2008). Empirical Evidence from Hangzhou's Urban Land Reform: Evolution, Structure, Constraints and Prospects. *Habitat International*, 32(4):219–233.

Queiroz, R. L. C., & Lago, L. C. D. (1995). Restructuring in Large Brazilian Cities: The Centre/Periphery Model. *International Journal of Urban & Regional Research*, 19(3):369–382.

Rakodi, C. (1998). *Review of the Poverty Relevance of the Peri-Urban Interface Production System*. Research Report for the DFID Natural Resources Systems Research Program, London: Department for International Development.

Ribeiro, L. C. Q., & Correa do Lago, L. (1995). Restructuring in Large Brazilian Cities: The Centre/Periphery Model. *International Journal of Urban and Regional Research*,19(3):369–382.

Rondinelli, D. A. (1983).*Secondary Cities in Developing Countries*. Beverly Hills: Sage Publications.

Rondinelli, D. A. (1985).*Applied Methods of Regional analysis: The Spatial Dimensions of Development Policy*. Boulder, CO:Westview.

Satterthwaite, D., & Tacoli, C. (2003).*The Urban Part of Rural Development: The Role of Small and Intermediate Urban Centres in Rural and Regional Development and Poverty Reduction*. London: IIED.

Seto, K. C., & Fragkias, M. (2005). Quantifying Spatiotemporal Patterns of Urban Land-Use Change in Four Cities of China with Time Series Landscape Metrics. *Landscape Ecology*, 20(7):871–888.

Shi, L., Shao, G., Cui, S., X.L., Lin, T., Yin, K., & Zhao, J. (2009). Urban Three-Dimensional Expansion and Its Driving Forces – A Case Study of Shanghai, China. *Chinese Geographical Science*, 19(4):291–298.

Sieverts, T. (1999). *Zwischenstadt*. Wiesbaden:De Gruyter.

Stöhr, W. B. (1981). Development from Above or Below?: The Dialectics of Regional Planning in Developing Countries. *Urban Studies*, 19(4):430–432.

Sui, D. Z., & Zeng, H. (2000). Modelling the Dynamics of Landscape Structure in Asia's Emerging Desakota Regions: A Case Study in Shenzhen. *Landscape and Urban Planning*, 758:1–16.

Tacoli, C. (2004). Rural-Urban Linkages and Pro-Poor Agricultural Growth: An Overview. OECD DACPOVNET Agriculture and Pro-Poor Growth Task Team Helsinki Workshop:2–3.

Tan, M., Li, X., Xie, H., & Lu, C. (2005). Urban Land Expansion and Arable Land Loss in China – A Case Study of Beijing-Tianjin-Hebei Region. *Land Use Policy*, 22 (3):187–196.

Thomas, D. (1974). The Urban Fringe: Approaches and Attitudes. In Johnson,J. H. (Ed.) *Suburban Growth, Geographical Process at the Edge of the Western City*, Aberdeen: Aberdeen University Press.

Tian, L. (2015). Land Use Dynamics Driven by Rural Industrialization and Land Finance in the Peri-urban Areas of China: The Examples of Jiangyin and Shunde, *Land Use Policy*, 45:117–127.

Tian, L., Ge, B. Q., & Li, Y. F. (2014). Research on Temporal and Spatial Characteristics and Influencing Factors of Land Use Change in Peri-Urbanized Areas of Shanghai since the 1990s. *Urban Planning*, 6:45–51. (In Chinese).

Tian, L., Ge, B. Q., & Li, Y. F. (2017). Impacts of State-led and Bottom-up Urbanization on Land Use Change in the Peri-Urban Areas of Shanghai: Planned Growth or Uncontrolled Sprawl?*Cities*, 60(B):476–486.

Tian, L. (2014). *Land Values, Property Rights and Urban Development: Betterment and Compensation under the Land Use Rights of China*. London: Edward Elgar Publishing.

Tian, L.,&Liang, Y. L. (2013). The Industrialization and Land Use in Peri-Urban Areas: An Analysis Based on the Development of Three Top 100 County Economies in Three Regions. *Urban Planning Forum*, 34(5):30–37.

Tian, L., Yang, P. R., Dong, H. P., & Liu, Y. (2011). Financial Crisis and Sustainable Development Background Comparison and Enlightenment of Sino-US Urban Regulation Education Orientation. *International Urban Planning*, 2:99–105.

Unwin, T. (1989). Urban-rural Interaction in Developing Countries: A Theoretical Perspective. In Potter, R. B., & Unwin, T. (Eds.) *The Geography of Urban-Rural Interaction in Developing Countries*, London: Routledge Press.

Webber, M. M. (1964). The Urban Place and the Nonplace Urban Realm. In Webber, M. M. (Ed.) *Exploration into Urban Structure*. Philadelphia: University of Pennsylvania Press, 79–153.

Webster, C.& Lai, L. W. C. (2003). *Property Rights, Planning and Markets: Managing Spontaneous Cities*. Cheltenham: Edward Elgar.

Webster, D., & Muller, L. (2002). *Challenges of Peri-urbanization in the Lower Yangtze Region: The Case of the Hangzhou-Ningbo Corridor*. Working Paper, Asia/Pacific Research Center, Stanford: Stanford University.

Wei, Y. D., & Ye, X. Y. (2014). Urbanization, Urban Land Expansion and Environmental Change in China. *Stochastic Environmental Research and Risk Assessment*, 28(4):757–765.

Wissink, G. A. (1962). *American Cities in Perspective, with Special Reference to the Development of their Fringe Areas*. Assen: Van Gorcum.

World Bank. (2000). *Rural-Urban Linkages and Interactions: Synthesis of Issues*. Conclusions and Priority Opportunities Emerging from the 9 March Workshop. New York.

Wu, F., Zhang, F., & Webster, C. (2013). Informality and the Development and Demolition of Urban Villages in the Chinese Peri-Urban Area. *Urban Studies*, 50 (10):1919–1934.

Wu, F. L., & He, S. J. (2005). Changes in Traditional Urban Areas and Impacts of Urban Redevelopment – A Case Study of Three Neighbor-Hoods in Nanjing, China. *Tijdschrift Voor Economische en Sociale Geografie*, 96(1):75–95.

Wu, K. Y., & Zhang, H. (2012). Land Use Dynamics, Construction Land Expansion Patterns, and Driving Forces Analysis of the Fast-Growing Hangzhou Metropolitan Area, Eastern China (1978–2008). *Applied Geography*, 34, 137–145.

Xiao, P., Wang, X., Zhang, X., Feng, X., & Yang, Y. (2014). Detecting China's Urban Expansion Over the Past Three Decades Using Nighttime Light Data. *IEEE Journal of Selected Topics in Applied Earth Observations and Remote Sensing*, 7 (10):4095–4106.

Yang, Q. Q., Wang, K. L., & Yue, Y. M. (2009). Spatial Distribution of Rocky Desertification and Its Difference between Scales in Northwest Guangxi. *Acta Ecologica Sinica*, 29(7):3629–3640.

Yeh, A. G. O., & Wu, F. L. (1996). The New Land Development Process and Urban Development in Chinese Cities. *International Journal of Urban and Regional Research*, 20(2):330–353.

Yin, J., Yin, Z., Zhong, H., Xu, S., Hu, X., Wang, J., &Wu, J. (2011). Monitoring Urban Expansion and Land Use/Land Cover Changes of Shanghai Metropolitan

Area during the Transitional Economy (1979–2009) in China. *Environmental Monitoring and Assessment*, 177(1–4):609–621.

Yu, X. J., & Ng, C. N. (2007). Spatial and Temporal Dynamics of Urban Sprawl along Two Urban-Rural Transects: A Case Study of Guangzhou, China. *Landscape and Urban Planning*, 79(1):96–109.

Yue, W., Liu, Y., & Fan, P. (2010). Polycentric Urban Development: The Case of Hangzhou. *Environment and Planning A*, 42(3):563–577.

Zhang, L., Le Gates, R., & Zhao, M. (2016). *Understanding China's Urbanization: The Great Demographic, Spatial, Economic, and Social Transformation*. London: Edward Elgar Publishing.

Zhao, M., Wu, B. H., &Yuan, S. Q. (2010). Research into the Spatial Structure of Vacation Destination in Suburban Areas of Beijing City: Based on the Change of Urban Center Distance. *Journal of Chongqing Normal University*, 27(1):74–78.

Zhao, P. J. (2012). Urban-Rural Transition in China's Metropolises: New Trends in Peri-Urbanization in Beijing. *International Development Planning Review*, 34(3):269–294.

Zhou, Z. H. (2014). Towards Collaborative Approach? Investigating the Regeneration of Urban Village in Guangzhou, China. *Habitat International*, 44:297–305.

Zhu, J. M. (2004).From Land Use Right to Land Development Right: Institutional Change in China's Urban Development. *Urban Studies*, 41(7):1249–1267.

Zhu, J. M. (2005). A Transitional Institution for the Emerging Land Market in Urban China. *Urban Studies*, 42(8):1369–1390.

Zhu, J. M. (2013). Governance Over Land Development during Rapid Urbanization under Institutional Uncertainty, With Reference to Peri-Urbanization in Guangzhou Metropolitan Region, China. *Environment and Planning C: Government and Policy*, 31(2):257–275.

Zhu, J. M. (2017). Making Urbanization Compact and Equal: Integrating Rural Villages into Urban Communities in Kunshan, China. *Urban Studies*, 54(10):2268–2284.

Zhu, J. M., & Guo, Y. (2014) Fragmented Peri-urbanisation Led by Autonomous Village Development under Informal Institution in High-density Regions: The Case of Nanhai, China. *Urban Studies*, 51(6):1120–1145.

Zhu, J. M., & Guo, Y. (2015).Rural Development Led by Autonomous Village Land Cooperatives: Its Impact on Sustainable China's Urbanization in High-Density Regions. *Urban Studies*, 52(8):1120–1413.

Zhu, J. M., & Hu, T. (2009). Disordered Land-Rent Competition in China's Peri-Urbanization: Case Study of Beiqijia Township, Beijing. *Environment and Planning A*, 41(7):1629–1646.

3 Evolution of urban–rural relationship and peri-urban areas development in China

This chapter goes through the evolution of the urban–rural relationship since the establishment of the People's Republic of China in 1949 in order to provide a backdrop to understand "dual track" urbanization: state-led and bottom-up urbanization. We select the peri-urban areas of Shanghai in the YRD, Guangzhou in the PRD, and Beijing in the BTH as case studies, examine their social-economic changes since the 1990s, and analyze the socio-economic characteristics of peri-urban areas.

3.1 An overview of urban–rural relationship in China

The urban–rural relationship is one of the most important factors to social and economic development in contemporary China. Over the decades, this relationship has experienced a tortuous evolutionary process.

During the Mao era (1949–1977), a strong priority was placed upon industry and urban-biased institutional arrangements, leading to a large disparity between urban and rural residents. The flexible driving channel of geographical and social mobility had been gradually shackled by top-down control. In other words, the urban–rural dual structure came into being during this period and the urban and rural landscapes were artificially divided into isolated worlds.

Since the reform opening (1978–), the deepening market economy reform has broken the sealed channel; meanwhile, the actual driving force of urban–rural relationship had shifted from state-dominant to the mixed forces of state and market. Although urban–rural disparity has been decreasing since the 2000s, the dual structure exists. How to achieve the goal of urban–rural coordinated development is a key issue for future sustainable development in China.

3.2 Urban–rural divide from 1949 to 1978

3.2.1 Barriers to urban–rural migration from 1949 to 1952

With the establishment of the People's Republic of China, the opening and harmonious urban–rural relationship that existed since feudal society remained, and there were few institutional or geographic barriers between the urban and rural populations. Peasants could go to cities freely, and urban residents could move to the countryside. Meanwhile, in order to reconstruct the national economy which was destroyed by war, the Communist Party of China (CPC) released the control of urban–rural flow. In reality, some development plans were made to accelerate the urban–rural goods exchange and reduce the urban–rural disparity.

First, an urban–rural commodity market was established, and the central government took various measures to bring agricultural products to urban areas, and to sell industrial products into rural areas. As a result, peasants received higher incomes, industrial enterprises received adequate agricultural products, and privately owned industrial and commercial enterprises flourished. This expanded economic and social exchanges between urban and rural areas. Second, the prices for agricultural products rose four times from 1951 to 1952, which narrowed the gap between prices of industrial and agricultural goods. Third, in the summer of 1950, the central government published the Agrarian Reform Law, which abrogated landownership of the landlord class and introduced peasant landownership. During the next two years, about 47 million hectares of land was confiscated from landlords and redistributed to nearly 310 million peasants who had little or no land (Wan, 2008). The ideal of "land to the tiller" was gradually realized. In other words, the land reform movement emancipated the productive forces, increased agricultural productivity, and became the cornerstone of industrialization in China.

From 1949 to 1952, the transient harmonious urban–rural relationship accelerated the reconstruction speed of national economy. Industrial and agricultural developments were both very prosperous in China. For example, the output of steel and pig iron increased nearly seven times during this period, and the output of grain in 1952 topped the highest level in history. The peasant households' income increased more than 30%. The urbanization rate increased from 10.64% in 1949 to 12.46% in 1952, and around 300 million peasants moved from rural areas to cities (*Source*: China Statistics Yearbook, 1980–2017).

3.2.2 The formation of the urban–rural dual structure (1953–1978)

Although the national economy recovered rapidly in the early days of the new China, backward productivity remained a big problem for the state. Industrial output accounted only for 25.3% of total GDP, and industrial employment accounted only for 6.0% of total employment in 1952 (Han, 2009). In order to push forward the industrialization process, the central committee decided to initiate large-scale industrial development. According to the First Five-year Plan (1953–1957), heavy industry was listed as the development priority of China, and 156 giant heavy industry projects and countless large and medium-sized projects were at the top of the agenda of government. As a result, a contradiction between industrial demand and agricultural supply emerged, and it had a significant negative influence on the urban–rural relationship. In the following years, the establishment of a unified purchase and sale system, the household registration system, and the people's commune system began to restrict the movement of peasants and curb the price of agricultural goods, and an urban–rural dual structure gradually formed.

In order to deal with the food shortage, a unified purchase and sale system was established in 1953 to withdraw the free market for agricultural products and required farmers to sell the majority of surplus grain to the state at very low state-imposed prices. At the same time, food supply was under strict control. The state forbid free business by private merchants and issued a food rationing system in which urban families were given fixed food coupons. Obviously, the unified purchase and sale system played an important role in easing the tension of food supply and marketing, but widened the gap between urban and rural areas. First, almost all agriculture production flowed into the market by top-down control. The peasants had completely lost their rights as producers and operators. Second, the agricultural surplus value was deprived by this system, which severely affected the production enthusiasm of peasants.

Meanwhile, the open relationship between urban and rural in the recovery period of the national economy brought many unemployed migrants to the city. To solve this problem, the Chinese State Council issued a directive to dissuade peasants blindly migrating to the city. However, this did not work because of the increasing urban–rural disparities. The influx of peasants into cities escalated and began to be a serious burden, which had a severe negative effect on urban development. Therefore, the CPC Central Committee and State Council continuously issued a number of directives of prohibiting the peasants' freedom to leave their rural homes. On the local level, a series of drastic measures had been taken in some cities. For instance, the city's food sector could not supply any food to unemployed peasants and industrial and mining enterprises could not employ workers from rural areas without authorization. In January 1958, the restriction on migration culminated in the implementation of China's first household registration (*hukou*) system that has long separated people into two groups, either urban (non-agricultural) or rural (agricultural) residents according to where they were born. Individuals were allowed to move downward (to other lower level cities or to the countryside) or parallel, but not upward. It was extraordinarily difficult for rural residents to migrate to urban areas unless they were admitted by a university or married an urban resident. Since then, a distinct demographic barrier has been set up between urban and rural areas. Peasants were denied access to abundant food supplies, higher wages, and welfare benefits which could be found in the cities. In particular, the household registration system made urban and rural residents politically unequal.

While the household registration system and the unified purchase and sale system had kept peasants out of the city gate, the people commune system had locked the peasants firmly in the farmland. The traditional Chinese rural area has the right of autonomy, and the state's rights can only reach the county level directly. The state never effectively controlled the countryside until the formation of the people commune system in the late 1950s. As a grass-roots political organization in China, the commune was based on the national interest, taking public ownership as the foundation, and fully controlled the two basic factors of production in the rural area: land and labor. All peasants had to give up most of their private property to the commune. They also had lost the autonomy that should had been their own, such as the right to control their land and production, freedom of choice, free movement and the right to trade goods, and so on. In communal society, the only source of income for most rural families was work-point (*gongfen* in Chinese), and the peasants who belong to the commune always got up early and worked hard in order to get more work-points. Li Kelin, the former head of the agriculture department of "People's Daily," once said, "Which peasant wasn't a commune member? Which peasant didn't rely on the work-point (Ma, 2008)?" In order to consolidate the collective economy, different regional communes set various control regulations to restrict the freedom of peasants. For example, if someone wanted to pay a visit to relatives and friends, he or she should be approved by the commune. If artisans wanted to work, he or she should be approved by the commune and pay some management fees. From 1958 to 1978, the people commune system further enhanced the urban–rural dual structure.

In the early 1960s, the CPC central committee noted the differences between urban and rural areas and stated as their goals in the Third Five-year Plan to, "vigorously develop agriculture and solve the problem of food and clothing." However, the Cultural Revolution erupted in 1966, throwing China into the midst of unprecedented turmoil. This not only hindered the implementation of the Third Five-year Plan (1966–1970), but also made rigid urban and rural relations worse. After 1968, Chairman Mao proposed that it was necessary for intellectuals to go into the countryside and receive re-education in camps. The Down to Countryside Movement was fully under way. It is estimated that about 16 million government officials and intellectuals moved from the city to rural areas, and their identities were changed into rural *hukou* (Han, 2009). This unusual population flow made countless city families stay separated for many years, and large numbers of the surplus labor force, even elite labor, had to stay in rural areas for many years.

To summarize, the combination of the unified purchase and sale system, household registration system, and the people's commune system were the root of the urban–rural dual structure in China from 1952 to 1978. It resulted in various political, economic, and social problems. For instance, the countryside lagged far behind the city in terms of social and economic development and the gap between industry and agriculture increased. The crucial contradictions between land and peasants became increasingly acute. The share of the agricultural sector in GDP seriously fell, and the share of industry, especially heavy industry, had rapidly rose from 1952 to 1978. The Index of Industrial Output Value was 100 in 1952 and increased to 1,734.4 in 1979, with a heavy industry index of 2,991.6, compared with 249.4 of agriculture during the same period. In terms of peasants' income, the net income of peasants per capita was 133.57 Yuan, only a third of that of urban residents. The consumption level of peasants increased by 57.5% from 1952 to 1978, only half the growth rate of urban residents for the same period (Bai, 2012). Most countries use dual structure intensity[1] as an index to measure urban–rural development disparities. According to a statistical analysis by Kuznets, the maximum value of dual structure intensity was 4.09 times in other developing countries. However, it was as high as 6.08 times in 1978 China (Zhang, 2001). Faced with the increasing disparity, the CPC central committee had to adopt policies aimed at alleviating this disparity while promoting national economic development after the end of the Cultural Revolution.

3.3 Urban–rural interaction since the reform opening

The pursuit of an extreme egalitarian society under the planned economy brought severe social and economic costs (Whyte, 2010).After the Cultural Revolution, poverty and chaos were prevalent throughout all of China; industrial and agricultural production was inefficient, the fiscal deficit was severe, the supply of materials was scarce, and the GDP per capita was only 9.1% of the world average (Tian & Liang, 2013). In response to this situation, the Third Plenary Session of the 11th Central Committee of the Chinese Communist Party, held in December 1978, initiated the prelude to reform and opening-up. Afterward, the planned economy began to transform into a market-oriented economy in regard to competition and regulations. This change had a significant impact on the urban–rural dual structure.

In 1978, the reform and opening-up policies were launched, reflecting Chinese leadership's shift of focus from ideological thought to economic growth (Knight et al., 2006).After 1978, China's urbanization developed rapidly, with a growing number of cities and urban population. Meanwhile, rural enterprises for non-agricultural industries began to appear (Ho & Lin, 2003), which partly altered the traditional urban–rural labor division. The urban–rural interaction evolved constantly under the impact of public policies and social transformation and can be roughly divided into two phases (Figure 3.1):

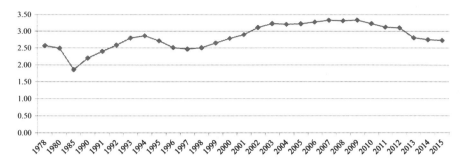

Figure 3.1 Urban–rural income ratio in China since 1978
Source: China Statistical Yearbook (1980–2017)

3.3.1 Urban bias period (1978–2003)

Facing the rigid relationship between urban and rural areas and the stagnation of rural production, Deng Xiaoping, the commander of the reform and opening-up, decided that the economy reform should first start in rural areas. Deng pointed out that 80% of China's population lived in rural areas, and the society would not be stable if the state did not solve the living problems of these 80%. He also addressed that the industrial development, commerce, and other economic activities could never run well with this 80% rural population in poverty (Literature Research Office of the CPC Central Committee, 1984). With the economy structure reform taking place in rural areas first, the transformation of urban–rural relations in China had been initiated.

From 1978 to 1985, the urban–rural gap in China narrowed to some extent as a result of the social and economic reforms taking place in rural China (Knight et al., 2006). The reforms were first launched at the grassroots level and then supported by the Chinese leadership with a series of new policies. On the one hand, the close of the people's communes and the establishment of the household contract responsibility system shifted rural land use from collectives to households (Ho & Lin, 2003), which greatly increased peasants' incentive to produce and resulted in more efficient use of farmland. Moreover, in December 1982, the Constitution of the People's Republic of China stipulated that most land in rural areas and suburbs shall belong to the collectives and land for homestead and private plots shall belong to the collective ownership. Thus, the rights of peasants to household contract land management were legally guaranteed and the rural productive forces were effectively liberated. On the other hand, the previous way of the communes such as enforced procurement of agricultural products was gradually replaced by peasants' autonomous management. The government decided to reduce the types and amounts of monopolistic farm products to give peasants more rights to independent management.

In January 1985, the Central Committee of the Communist Party stipulated ten policies on the rural economy, including that the state should no longer assign monopoly purchase and marketing tasks to peasants' agricultural products, and adopted new methods of contract orders and market purchases according to different circumstances. After these new polices were in place, the monopolistic managing system of farm products was withdrawn from the stage of history, with farmers completely regaining the land and their initiative of managing their agricultural products. Moreover, the pricing and trading system for farm products was restructured, leading directly to arise in food prices. In 1978, the Third Plenary Session of the Eleventh Central Committee decided to increase the purchase price of foodstuffs by 20% and the price of other farm products such as cotton oil also increased to varying degrees. As of 1984, the purchase price of agricultural and sideline products had also increased 53.6% compared with 1978 prices. These dual incentives stimulated the rapid development of agriculture production (Xia, 2008). Between 1978 and 1984, peasants' average net incomes increased by 165.9% from 133.6 to 355.3 Yuan, while the urban–rural income ratio decreased from 2.57:1 to 1.71:1 (*Source*: China Statistics Yearbook, 2000), indicating an alleviation of urban–rural disparity in this period.

The emphasis on rural areas did not last long. The Chinese government decided to shift the emphasis of their reform policies from rural to urban in the mid-1980s (Knight et al., 2006), which once again deepened the urban–rural divide in China. The promotion of the opening policy and market coordination accelerated economic reform in urban areas with rapid growth by 11.7% in urban incomes from 1984 to 2002 (Whyte, 2010). At the same time, the neglect of rural reform in this period led to a stagnation in agricultural development. In 2002, the urban–rural income ratio expanded to 3.11:1, along with an apparent variation in other unmeasured aspects such as access to education, healthcare, and quality housing. In 2000, urban residents were reimbursed for medical fees at a 40% rate of total expenses and rural residents only at a 10% rate. In 2002, over 90% of the eligible urban population enjoyed the minimum living allowance, compared with only 25% in rural areas, reflecting a widening gap between urban and rural living standards (Knight et al., 2006). Between1984 and 2002, the imbalance between the reforms and developments of urban and rural areas was enlarged, leading to a series of social problems.

3.3.2 The urban–rural coordinated development strategy since 2003

With regard to the expanding urban–rural gap and relevant social problems, the Chinese government proposed the urban–rural coordinated development strategy in October, 2003, giving priorities to agriculture, rural development, and farmers' well-being. In January 2006, the "People's Republic of China Agricultural Tax Regulations" was officially abolished, and the agricultural tax that had existed since ancient times was finally abolished. After the tax reform, the total economic burden of peasants across the country was reduced by 1250 billion Yuan annually and the per capita burden reduced by about 140 Yuan.[2] The suspension of the agricultural tax also increased peasants' civil rights. After the tax reform, governments increased the budget of public services in rural areas, which not only improved the subsidies for peasants, but also increased rural infrastructure investment. The cancellation of tax on farm products also reflected the development strategy of "industry re-feeding agriculture and urban areas re-feeding rural areas," which reduced the burden of Chinese peasants and rural governments and increased peasants' incomes. Even after the tax reform, the huge disparity between urban and rural areas remained in place before the coordinated development strategy could gain significant results. Thus, for a long time, cities still excelled over rural areas in how quickly they developed.

The proposal for constructing beautiful countryside put forward by President Xi in 2013 further promoted urban–rural integrative development. Since 2013, the "Beautiful Village Plan" has become the focus of the overall development of urban and rural areas throughout China and various types of poverty alleviation funds have been started. The "Beautiful Village Plan" proposal not only effectively improved the built-up environment and public infrastructure in rural areas, but also brought about a steady increase in the income of peasants as well as rural tourism and urban–rural exchanges. The construction of communal facilities and the transformation of the economic structure in rural areas contributed to the overall improvement of peasants' living standards including incomes and access to public welfare and social services. In 2016, the urban–rural income ratio decreased to 2.72:1, indicating the potential for dismantling the urban–rural duality (*Source*: China Statistics Yearbook, 1980–2017).

Moreover, the production and living standards of the population which migrated from rural to urban areas were also effectively improved. In July 2014, the state promulgated the regulation to provide basic public services to urban migrants so that they could enjoy equal rights in education, medical care, and registered urban population. In addition, a number of safeguards were also put forward for the legal rights and interests of the migrants who have moved from rural areas, such as establishing a market for property rights transfer in rural areas, protecting land contractual management rights, and ensuring various benefits such as education, medical treatment, and insurance and pension for rural migrants. In reality, however, the gap between urban and rural China still remains today, and migrants face with the institutional barriers caused by *hukou* and do not really enjoy the same welfare with urban *hukou* residents. Due to the lack of financial policy from the central government, local governments are not motivated to provide the same welfare to rural-to-urban migrants. Although some achievements have been made, there is a long way to go to fulfill the objectives of urban–rural coordinated development.

3.4 Socio-economic development of Peri-urban areas: case studies of Peri-urban areas in three urban agglomerations

This section is devoted to discussing the socio-economic development in three most dynamic urban agglomerations in China: the Yangtze River

Delta (YRD), Pearl River Delta (PRD), and the Beijing, Tianjin and Hebei region (BTH), by taking three typical peri-urban areas: Jiangyin and Kunshan in the YRD, Shunde and Nanhai in the PRD, and the Shunyi in the BTH Urban Agglomeration (Figure 3.2).

3.4.1 An overview of development of three urban agglomerations since the reform opening

Up to now, the PRD, YRD, and BTH regions have grown into the most dynamic regions in China, boasting the fastest social and economic development and the highest urbanization level in China. The social and economic development in the three urban agglomerations is apparently path-dependent. As a result, their socio-economic development and land-use models vary with the different development paths they have chosen (Tian & Liang, 2013).

The PRD is a forerunner of economic reform and opening-up to the outside world in China and one of the most important economic centers of China. The

Figure 3.2 Locations of three urban agglomerations and cases of peri-urban areas
Source: drawn by authors

PRD plays a significant and strategic role in driving social and economic development in neighboring areas and overall reform in China. In 1985, the Central Government of China decided to establish the PRD Economic Development Zone. In 1988, Guangdong Province was designated as a Pilot Zone of Economic System Reform. From 1990 to 2014, the permanent population of the Delta increased from 23,506,400 to 57,633,800, accounting for 4.2% of the total population of China. Subsequently, by 2014, the urbanization rate of the Delta had risen to 84.12%. During the same period, the GDP of the Delta grew from 100.688 billion Yuan to 5,783.3 billion Yuan, accounting for 9.1% of the total GDP of China, with the GDP per capita jumping to over the 10,000 Yuan. The industrial structure was 1.9%:44.9%:53.2% (Primary, Secondary, and Tertiary) in 2014, with mainly the labor-intensive industries, i.e., garment, toy, and home appliance industries (*Source*: Urban Agglomeration Statistics Yearbook of PRD, 2015). In recent years, driven by high-tech industries, the Delta has become one of the largest producers and exporters of consumer electronics and other consumer goods.

With an advantageous geographic location and favorable state policies, the YRD seized the opportunity of the Pudong New District Development initiated in Shanghai in the early 1990s and moved into a rapid industrialization and urbanization stage. From 1990 to 2014, the permanent population[3] of the YRD surged from 123,389,000 to 158,937,400, accounting for 11.0% of the total population of China. By 2014, 67.04% of the Delta had been urbanized, one of the highest rates in China. Meanwhile, the GDP of the Delta had grown from RMB 310.36 billion in 1990 to RMB 12,882.91 billion, with GDP per capita increasing to RMB 81,069. This continuous economic growth gave rise to a continuous upgrading of the local economic structure. Over two decades, the Delta had shifted its industrial structure from 50%:25%:20% in 1990 to 48.0%:46.2%:5.8% (Secondary, Tertiary, and Primary) in 2013 (*Source*: City Statistics Yearbook of YRD, 2015). It is also notable that this was a typical "2–3–1" structure. Driven by rapid growth in the secondary industry, particularly the manufacturing industry, the YRD economy has been growing rapidly. As an economic highland or economic engine of China, the YRD has become one of the six internationally recognized top urban agglomerations in the world.

Being the latest entrant of three urban agglomerations, the journey of the BTH began with the establishment of the Tianjin Economic-Technological Development Area (TEDA) in 2006. After 2006, this region began to enjoy more favorable policies and greater autonomy for economic development (Wang & Ren, 2011). From 1990 to 2014, the permanent population in the BTH region increased from 80,670,000 to 111,000,000, accounting for 8.1% of total population of China. In 2014, the BTH region achieved a GDP of 6,647.45 billion Yuan, or 10.4% of China's total GDP, with its GDP per capita growing to 84,410 Yuan (*Source*: Statistics Yearbook of Beijing, Tianjin and Hebei Province, 2015). By 2014, the BTH had become the largest and most dynamic economic center in Northern China. In terms of industrial structure, the BTH is built on the basis of capital-intensive industries, particularly, the heavy chemical industry. Currently, it is the center of heavy chemical, equipment manufacturing, and high and new technology industries. By 2016, the proportion of the three industries of the BTH region was 5.2%:37.3%:57.5% (Primary, Secondary, and Tertiary).

3.4.2 Socio-economic development of peri-urban areas in the PRD: cases of Shunde and Nanhai

(1)   Study area

Shunde and Nanhai, located in the mid-south of Guangdong Province, are two districts of Foshan municipality, which is the hinterland of the Pearl River Delta. The two districts lie to the west of Guangzhou. After the reform and opening-up in 1978, Shunde and Nanhai utilized the geographical advantages of neighboring Hong Kong and Macao and national preferential policies to make great achievements in rural industrialization and urbanization. Both districts were converted from counties to cities in 1992 and became districts of Foshan municipality in 2003.

The total land area of Nanhai is 1,073.82 km^2 while Shunde is 806.57 km^2. Nanhai had a *hukou* population of 1.28 million while Shunde had 1.278 million in 2015. The regional GDP was 2,228.99 billion Yuan in Nanhai while the GDP was 2,586.69 billion Yuan in Shunde, accounting for 27.85% and 32.32% of the GDP in Foshan municipality, respectively (*Source*: Foshan Statistics Yearbook, 2016).

(2)   Economic development and industrial upgrading

After the reform opening, Shunde and Nanhai, making use of their advantageous location and preferential policies, vigorously promoted the development of their commodities markets and rural enterprises. By the early 1990s, they achieved rapid growth and transferred from traditional agricultural counties to newly emerging industrial counties. As can be seen from Figure 3.3, Shunde and Nanhai have had experienced several stages since the 1990s, which can be divided into three stages:

Stage 1: rapid growth (1992–1997)

Shunde and Nanhai maintained a high rate of economic growth during this period. In Shunde, the GDP increased 26 times from 4.75 billion Yuan in 1978 to 128.7 billion Yuan in 1994. The GDP growth of Shunde and Nanhai reached an astonishing rate of 46.4% and 45.6% in 1993.After 1994, economic growth slowed down compared with the initial developing period in both districts. With the 1997 Asian financial crisis, the GDP growth rate of Shunde and Nanhai fell to 10.4% and 10.2% in 1999 (*Source*: Foshan Statistics Yearbook, 2000)

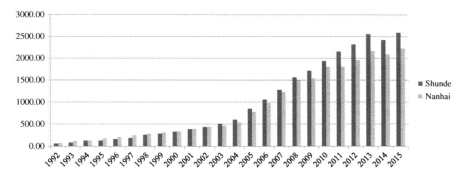

Figure 3.3 Changes of GDP of Shunde and Nanhai since 1992 (unit: 100 million Yuan)
Source: Foshan Statistics Yearbook (2016)

Stage 2: steady growth (1998–2005)

Between1998 and 2005, while gradually recovering from the effects of the Asian financial crisis, Shunde and Nanhai maintained a steady and relatively fast economic growth rate, with a GDP growth rate remaining above 10% and continuously increasing. The annual GDP growth rate of Shunde reached a new peak of 41.35%, and the local GDP was 849.62 billion Yuan in 2005. Nanhai's GDP growth rate also crested at 22.3% in 2006, and its local GDP reached 980 billion Yuan.

In this period, the economy grew at a slower pace in Nanhai than in Shunde. Figure.3.12 describes the changes of GDP in these two areas. As can be seen, Nanhai previously exceeded Shunde in economic aggregate in the 20th century. In 1999, the GDP was 305 billion Yuan in Nanhai and 286.3 billion Yuan in Shunde. However, Shunde surpassed Nanhai with a higher growth rate during this stage. In 2006, GDP of Shunde was 1,058.4 billion Yuan, nearly 80 billion more than that of Nanhai. Shunde's better economic performance is in part attributed to its reform in administrative and property systems after being selected as a demonstration area in institutional reform in Guangdong Province (Tian & Liang, 2013).

Stage 3: explosive rapid growth (since 2006)

From 2005 to 2015, the GDP in Shunde and Nanhai nearly tripled, from 849.62 to 2,587.45 billion Yuan in Shunde and from 771.89 to 2,226.97 billion Yuan in Nanhai. During this period, the GDP grew at an average growth rate of 13.0% in Shunde and 13.6% in Nanhai. In 2015, the growth rate dropped to 8.5%in both districts due to the impacts of global economy on the export-oriented economy in these two areas (*Source*: Foshan Statistics Yearbook, 2016).

During 1992–2015, the GDP per capita of Shunde increased by 15 times from 6,702 to 102,538 Yuan in Shunde, and GDP per capita increased by 11 times

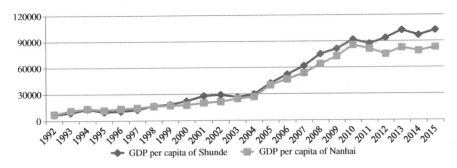

Figure 3.4 GDP per capita of Shunde and Nanhai since 1992 (unit: 100 million Yuan)
Source: Ibid.

from 7,256 to 82,961 Yuan in Nanhai. As Figure 3.6 shows, there is an overall increasing trend in terms of GDP per capita in both districts. After 2005, the growth rate of GDP per capita gradually declined in the two districts, and even fell in 2014 (*Source*: Foshan Statistics Yearbook, 2016). By comparison, the growth rate of GDP per capita was lower in Nanhai than that of Shunde on the whole. According to Figure 3.4, fluctuation trends in the GDP per capita growth rate were fairly similar in Shunde and Nanhai during 1992–2015. Although the growth rate was less stable in Shunde, it maintained a relatively high value between1997 and 2002 with an average rate of 19%, while it was 8% in Nanhai during the same period.

Figure 3.5 presents the change of industrial structure of Shunde and Nanhai, and we find that the secondary industry dominated the local economy, while the primary industry took up only a small percentage. In Shunde, the industrial structure changed from 6%:55%:39% (Primary Industry:Secondary Industry:Tertiary Industry) in 2000 to1.5%:58.5%:40% in 2015. From 2000 to 2008, the secondary industry developed rapidly, and its proportion rose from 55% to 64.8% and the share of tertiary industry fell from 39% to 33.2%. After 2008, service industries grew rapidly, and the proportion of tertiary industries increased to 40%, while the proportion of secondary industries dropped to 58.5% in 2015.

In 2015, the industrial structure in Nanhai was 2.2%:59.4%:38.4%. From 2007to 2012, along with development of the tertiary industry, the share of its secondary industry decreased to 43.8% in 2014, close to the share of the tertiary industry (43.1%). After 2012, the share of its secondary industry began to rise again.

(3)   Demographic and social development

As shown in Table 3.1, the *hukou* population both grew slowly in Shunde and Nanhai in last two decades. During 2002–2015, the *hukou* population increased

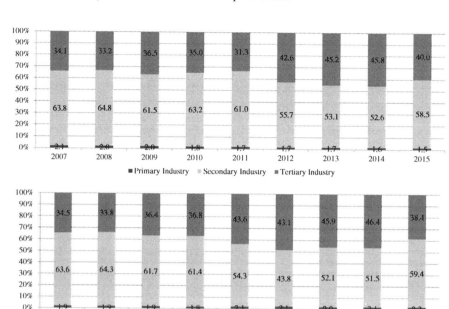

Figure 3.5 Industrial structure in Shunde (upper) and Nanhai (low) (2007–2015)
Source: Shunde and Nanhai Statistics Yearbook (2016)

by merely 175,300 from 1,109,600 to 1,284,900 in Shunde, with an annual average growth rate of 1.22%. During 2005–2015, the *hukou* population increased by merely 162,900 from 1,117,100 to 1,280,000 with an average growth rate of 1.46% in Nanhai. The migrant population accounts for around half of permanent population in Shunde and Nanhai. In 2015, the permanent population in Shunde was 2,535,300, while 49.32% of them, 1,250,400, was migrant population (Figure 3.6). In Nanhai, the percentage of migrant population in permanent population was even higher, taking up 54.21%.

Urban residents' income increased continually in the last decade in Shunde and Nanhai (Table 3.2). From 2005 to 2015, the disposable income of urban residents per capita doubled from 20,819 to 42,559 Yuan in Shunde, while it increased from 18,217 to 40,148 Yuan by 2.2 times in Nanhai. Compared with average level of Guangzhou province and China, disposable income of urban and rural residents is higher in Shunde and urban Nanhai. In 2015, disposable income per urban resident of Shunde and Nanhai was 1.36 and 1.22 times of nationwide average, 1.29 and 1.16 times of the average level of Guangdong Province.

Rural residents' income in Shunde and Nanhai also increased significantly in the last decade. In 2005, the average net income per rural

Table 3.1 Population growth in Shunde and Nanhai from 2005 to 2015 (unit: 1,000 persons)

	Shunde			Nanhai		
	Hukou Population	Migrant Population	Permanent Population	Hukou Population	Migrant Population	Permanent Population
2005	1,103	917	2,020	1,117	808	1,925
2006	1,160	869	2,029	1,130	964	2,094
2007	1,190	880	2,070	1,140	1,050	2,190
2008	1,190	880	2,070	1,159	1,173	2,332
2009	1,213	890	2,103	1,175	943	2,118
2010	1,220	890	2,110	1,189	965	2,154
2011	1,238	1,235	2,473	1,209	1,020	2,228
2012	1,248	1,236	2,483	1,225	1,008	2,233
2013	1,259	1,236	2,495	1,245	1,783	3,028
2014	1,271	1,231	2,502	1,265	1,976	3,241
2015	1,285	1,238	2,523	1,280	1,516	2,796

Source: Foshan Statistics Yearbook (2016)

Table 3.2 Disposable income of urban and rural residents in Shunde and Nanhai from 2005 to 2015 (unit: Yuan)

District	Shunde		Nanhai		Guangdong		Nationwide	
Per Capita Disposable Income	Urban	Rural	Urban	Rural	Urban	Rural	Urban	Rural
2005	20,819	9,331	182,17	8,744	14,770	4,691	10,493	3,255
2006	22,291	9,798	21,983	9,501	16,016	5,080	11,760	3,587
2007	28,345	10,637	24,495	10,359	17,699	5,624	13,786	4,140
2008	26,433	11,179	25,961	11,158	19,733	6,400	15,781	4,761
2009	29,215	11,850	28,309	12,326	21,575	6,907	17,175	5,153
2010	30,618	12,543	29,978	13,285	23,898	7,890	19,109	5,919
2011	34,262	14,148	32,295	14,805	26,897	9,372	21,810	6,997
2012	38,861	16,062	36,348	16,673	30,227	10,543	24,565	7,917
2013	42,749	18,111	39,843	18,516	33,090	11,669	26,955	8,896
2014	38,767	24,537	36,886	23,655	32,148	12,246	28,844	9,892
2015	42,559	26,860	40,148	25,909	34,757	13,360	31,195	11,422

Source: China, Guangdong province, Foshan Statistics Yearbook (2016)

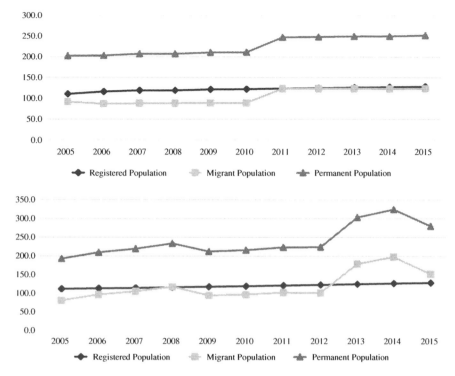

Figure 3.6 Growth of population in Shunde (upper) and Nanhai (low) from 2005 to 2015 (unit: 10,000 persons)
Source: Foshan Statistics Yearbook (2016)

resident was 8,744 and 9,331 Yuan in Nanhai and Shunde, respectively. By 2015, the income of rural resident increased by nearly three times in both areas, and reached at 25,909 Yuan in Nanhai and 26,860 Yuan in Shunde. During 2005–2015, the average net income per rural resident in Shunde and Nanhai was much higher than the average level of Guangdong province and nationwide. In 2015, the average net income per rural resident in Shunde and Nanhai was 2 and 1.9 times of the average of Guangdong (13,360 Yuan), while 2.4 and 2.3 times the nationwide average level (11,422 Yuan).

From 2005 to 2015, the income ratios of urban and rural residents were lower in Shunde and Nanhai than the average in Guangdong and nationwide, indicating that the urban–rural gap is smaller in Shunde and Nanhai than that in Guangdong Province and nationwide. In 2015, the national income ratio of urban–rural residents was 2.73, and the Guangdong ratio was 2.6; however, it was only 1.58 in Shunde, and 1.55 in Nanhai. Moreover, there is a decreasing tendency of the urban–rural income ratios in Shunde and Nanhai, and the relatively small urban–rural gap is mostly attributed to the development of local rural industries in the peri-urban areas.

3.4.3 Socio-economic development of peri-urban areas in the YRD: cases of Jiangyin and Kunshan

(1) Study area

Jiangyin is located in southern Jiangsu Province. It occupies a land area of 987.53 km^2 and had a permanent population of 1,637,000 in 2015. The urbanization rate reached 69.42% and a total GDP of 288 billion Yuan, making it the first-class county/county-level city economy in China for 10 consecutive years (*Source*: Kunshan Statistics Yearbook, 2016). Kunshan is another county-level city in southeast Jiangsu Province, occupying a land area of 927.68 km^2. In 2015, Kunshan had 1,651,200 permanent residents, urbanization rate was 72.04%, and GDP reached RMB 308.001 billion. It has been a first-class county/county-level city economy in China for 10 consecutive years (*Source*: Kunshan Statistics Yearbook, 2016).

(2)   Economic and industrial upgrading

In general, since the reform opening, the economic development of both cities can be divided into the following three stages (Figure 3.7): rapid growth, explosive growth, and moderate growth.

Stage 1: rapid growth (1992–2003)

Since 1992, given their favorable geographic locations, preferential state policies, strong foreign investment, and effective institutional innovations,

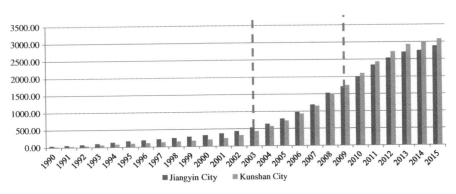

Figure 3.7 Changes of GDP of Jiangyin and Kunshan since 1990 (unit: 100 million Yuan)

Source: Jiangyin Statistics Yearbook (2016) and Kunshan Statistics Yearbook (2016)

industrialization process were accelerated in both cities. From 1992 to 2003, the GDP of Jiangyin and Kunshan increased 6 times and 10 times, respectively. Due to the dismantling township and village enterprises (TVEs), the GDP growth speed of Jiangyin fell from 39.87% to 12.28%, and that of Kunshan decreased from 62.20% to 14.20% from 1992 to 1996. Since then, the annual GDP growth rate in both cities ranged between 10% and 20% (*Source*: Jiangyin Statistics Yearbook, 2004 and Kunshan Statistics Yearbook, 2004).

Stage 2: explosive growth (2003–2009)

From 2003 to 2009, Jiangying and Kunshan witnessed a period of unprecedented economic growth. During this stage, Jiangyin was able to maintain its GDP growth rate at above 20% per year, and its GDP reached 171.3 billion Yuan by 2009. During the same period, GDP growth was as high as 26% in Kunshan. In 2009, it overtook Jiangyin for the first time and achieved a total GDP of RMB 175.1 billion, showing great potential for further development (*Source*: Jiangyin Statistics Yearbook, 2010 & Kunshan Statistics Yearbook, 2010).

Stage 3: moderate growth (2009–2015)

After the 2008 global financial crisis, both cities saw a dramatic slowdown in their GDP growth rates. After 2012, in both cities, GDP growth fell slightly again and stayed at a level a little below 10%. In 2014, GDP grew by 1.77% in Jiangyin and 2.63% in Kunshan, the lowest rates since the reform opening. Despite the slow growth, the total GDP reached 288.3 billion Yuan in Jiangyin and 308 billion Yuan in Kunshan in 2015 (*Source*: Jiangyin Statistics Yearbook, 2016 & Kunshan Statistics Yearbook, 2016). From 1990 to 2015, the GDP per capita increased from 3,968 Yuan to 232,140 Yuan in Jiangyin, and from 3,563 Yuan to 391,345 Yuan in Kunshan, while that of Kunshan grew higher than that of Jiangyin.

Moreover, both Jiangyin and Kunshan witnessed changes in their industrial structure. Although their economic structure has been dominated by secondary industries, the proportions of primary industry and secondary industry kept declining and the proportion of tertiary industry steadily grew in both cities. From 1990 to 2015, the industrial structure of Jiangyin changed from 15.1%:67.9%:17% to 1.6%:55%:43.4%, and that of Kunshan changed from 22.6%:56.5%:20.9% to 0.9%:55.1%:44% (Figure 3.8). The dominant secondary industry and underdeveloped tertiary industry imposed new challenges to the socio-economic development in these cities. On the one hand, environmental pollution has become a serious problem due to scattered industrial land. Moreover, limited land resources make it very difficult for local enterprise to expand their production scale. On the other hand, although the economy has been dynamic, the shortage of high-quality urban public services has made these cities less attractive for high-tech and talents who play an essential role in economic development.

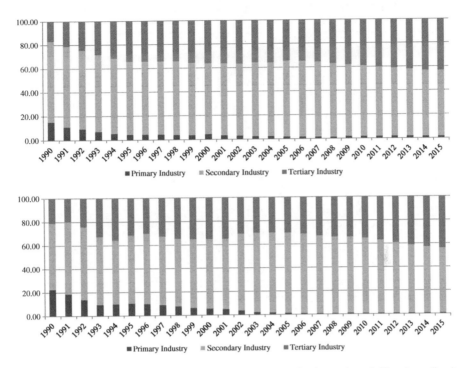

Figure 3.8 Change of industrial proportion of Jiangyin (upper) and Kunshan (low) since 1990

Source: Jiangyin Statistics Yearbook (2016) and Kunshan Statistics Yearbook (2016)

(3) Demographic and social development

Table 3.3 shows that *hukou* population growth was moderate in both cities. From 1990 to 2015, the population with *hukou* increased from 1,103,000 to 1,241,000 in Jiangyin, and from 564,600 to 787,000 in Kunshan. The population with *hukou* began declining in Jiangyin in 2013 while the amount of migrants increased. Jiangyin had 1,636,800 permanent residents, among which 395,800 was migrant population, and Kunshan had 1,651,200 permanent residents and 864,200 migrant population in 2015. The large number of migrant population put local social management and public service facilities under great pressure.

There was no obvious difference between the disposable income per urban resident and rural resident in Jiangyin and Kunshan. Both cities have maintained high growth in terms of disposable income per urban resident and rural resident and are higher than that of Jiangsu Province and the national average. By 2015, disposable income per urban resident increased to 50,701 Yuan in Jiangyin, and 50,749 Yuan in Kunshan, much higher than the provincial

Table 3.3 Population with *hukou* and permanent population in Jiangyin and Kunshan since 1990 (unit: 1,000 persons)

Year	Jiangyin		Kunshan	
	Population with hukou	Permanent Population	Population with hukou	Permanent Population
1990	1,103	–	564.6	–
1995	1,137.2	–	580.5	–
2000	1,151.8	–	594.6	–
2005	1,186.2	–	654.6	958.1
2010	1,207.1	1,594.8	711.3	1,646.3
2011	1,208.8	1,616.4	723.6	1,658.7
2012	1,212.6	1,624.3	737.6	1,638.9
2013	1,217.3	1,630	752.9	1,643.5
2014	1,232.1	1,634.7	769.7	1,650.3
2015	1,241	1,636.8	787	1,651.2

Source: Jiangyin Statistics Yearbook (2016) and Kunshan Statistics Yearbook (2016)

average level (37,174 Yuan) and national average level (31,195 Yuan). In terms of disposable income per rural resident, Jiangyin and Kunshan reached at 26,012 and 25,978 Yuan in 2015, 16,257 Yuan higher than the average of Jiangsu Province and more than twice the national average.

Comparing the incomes of urban and rural residents (Figure 3.9), we find that the expansion of the income gap mainly occurred from 2002 to 2009. In Jiangyin, the income gap increased from 1.49 to 2.06 and from 1.72 to 2.1 in Kunshan. After reaching its peak, the income differential began falling and then stabilized at around the level of 2.0. During the period from 2002 to 2009, Jiangyin experienced the fastest economic growth in its history and the income gap was notably expanded, but the income gap of Jiangyin was lower than the national average level and the Jiangsu Province average level. In 2015, the income difference of both cities remained at 1.92, lower than that of Jiangsu Province, 2.29, and the national level, 2.9. Compared with other regions, the rural economies were growing more vigorously due to the developed TVEs in both cities.

3.4.4 Socio-economic development of peri-urban areas in the BTH region: case of Shunyi

(1)   Study area

Located in the northeast corner of the Beijing Municipality, the Shunyi District is 30 km away from the Beijing city center. It occupies a land area of

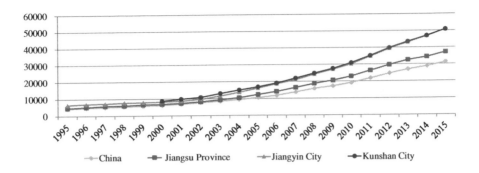

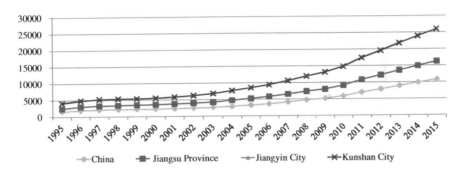

Figure 3.9 Comparison of disposable income per urban resident (upper) and rural resident (low) in Jiangyin and Kunshan since 1995 (unit: Yuan)

Source: Jiangyin Statistics Yearbook (2016) and Kunshan Statistics Yearbook (2016)

1,019.89 km² and hosted as many as 1.02 million permanent residents with a GDP of 144.09 billion Yuan in 2015. Shunyi had ranked as one of the top 100 county/county-level city economic powers in China for three consecutive years. In 1998, Shunyi was merged into one of the Beijing districts. Being the most powerful economy of all Beijing suburban counties, Shunyi is now a key link in Beijing's northeast development belt, and the core area of the Capital International Air Center.

(2) Economic and industrial development

After 1992, the economic development path of Shunyi can be divided into three development stages in Shunyi (Figure 3.10): first stage (1992–1999), second stage (2000–2009), and third stage (2009–2015). In the first stage, the total GDP of Shunyi increased from 2.486 billion Yuan in 1992 to 7.836 billion Yuan in 1999, growing at an average annual rate of 20%. In the second stage, from 2000 to 2009, GDP grew from 10.492 billion Yuan to 69.018 billion Yuan, with the highest growth occurring in 2000 (33.9%) and 2008 (52.73%). In the third stage,

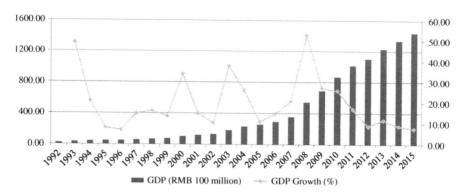

Figure 3.10 Changes of GDP and its growth of Shunyi since 1992
Source: Beijing Statistics Yearbook (2016)

GDP growth began to slow down. From 2012 to 2015, the rate generally floated around 10% and finally fell to 7.55% in 2015, the lowest in Shunyi's history since the 1990s. From 1992 to 2015, the GDP per capita increased by 30 times in Shunyi.

Figure 3.11 presents the change of industrial structure in Shunyi from 1992 to 2015. We can find that the proportion of the primary industry has been declining, the proportion of the tertiary industry has been steadily growing, and the proportion of the secondary industry rose for a long time period and then began to fall in 2008. From 1992 to 2008, economic development was dominated by secondary industry. The total structure shifted from 31.3%:53.9%:14.8% in 1992 to 3.7%:42.7%:53.6% in 2008. After 2008, the tertiary industry began taking up a bigger share in the structure. In 2015, the industrial structure was upgraded to

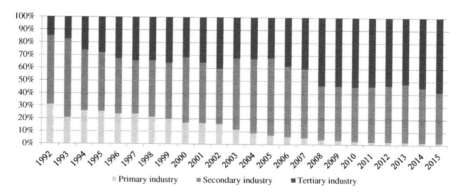

Figure 3.11 Change of the industrial structure in Shunyi since 1992
Source: Ibid

1.5%:40.1%:58.4%. The change of industrial structure and the development of tertiary industry have been closely related to the Capital International Air Center and its new industries (Tian & Liang, 2013).

(3) Demographic and social development

Since 1990, the population with *hukou* has been growing at a moderate rate in Shunyi. From 1990 to 2015, its population with *hukou* grew from 530,000 to 620,000. Meanwhile, due to the increase of migrant population, the permanent population grew much faster than *hukou* population. In 2015, the number of permanent population rose to 1,020,000, among which 402,000 were migrant population.

The disposable income of urban and rural residents had been rising rapidly in Shunyi (Figure 3.12). Compared with Beijing's average, the disposable income of urban residents was lower in Shunyi, but higher than that of the national average. In 2015, disposable income per urban resident reached 33,394 Yuan in Shunyi, lower than Beijing's average (52,859 Yuan) but higher than the national average (31,195 Yuan). In 2014, disposable income per

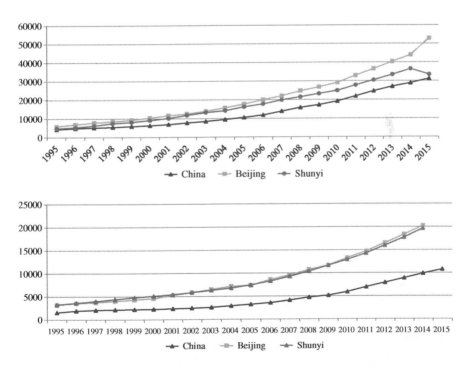

Figure 3.12 Comparison of disposable income per urban resident (upper) and rural resident (low) in Shunyi since 1995 (unit: Yuan)

Source: Beijing Statistics Yearbook (2016)

rural resident was 19,629 Yuan, slightly lower than the Beijing average (20,226 Yuan) but nearly twice as much as the national average (10,772 Yuan). Meanwhile, the disparity between the incomes of urban and rural residents increased from 1995 to 2005, and became smaller after 2005. The period from 2002 to 2008 was a time that witnessed the greatest income disparity between urban and rural residents (around 2:1). After 2008, the income ratio began narrowing and went down to 1.84:1 by 2014.

3.5 Summary

To summarize, the economy in peri-urban areas has been strong due to their advantageous location and developed manufacturing industries. With the co-existence of top-down investment and bottom-up rural industrialization, their urban–rural income disparity has been much smaller when compared with the regional and national average. Moreover, the developed economy in peri-urban areas has attracted a number of migrants who are beneficial for local economic growth. However, while migrants do provide some benefits, they also pose challenges for local governments in terms of the provision of public services.

As time has passed, the proportion of secondary industries has diminished in peri-urban areas, and the proportion of tertiary industries has grown. Secondary industries still play a dominant role in the case areas of the PRD and YRD regions, and tertiary industries exceed secondary industries and have become the leading industry in Shunyi. The difference in socio-economic development models in various regions has generated different impacts on the development pattern of peri-urban areas. In the PRD and YRD regions, bottom-up rural industrialization has been the main driving force behind social and economic development. However, in the BTH region, the influence of spill-over effects of the central city and top-down investments have been more significant than rural industrialization in local economic development. As a consequence, the land use of peri-urban areas has exhibited different characteristics in different regions, which will be discussed in the following chapter.

Notes

1 It is measured by income ratio of agriculture and non-agriculture populations.
2 www.chinaacc.com/new/253_268_201202/02ca517743065.shtml; accessed on 10/15/2015.
3 In China, permanent population is the sum of *hukou* residents and migrants who have resided in the city more than half a year.

References

Bai, Y. X. (2012). Urban–Rural Dual Structure Under the Chinese Perspective: Its Formation, Expansion and Path. *Academic Monthly*, 44(5):67–76.
Beijing Statistics Bureau. (2016) *Beijing Statistics Yearbook*. Beijing, China: China Statistics Press.

China Statistics Bureau. (1980–2017). *China Statistics Yearbook*. Beijing, China: China Statistics Press.

China Statistics Bureau. (2015). *City Statistics Yearbook of YRD*. Beijing, China: China Statistics Press.

Foshan Statistics Bureau. (2000, 2016). *Foshan Statistics Yearbook*. Beijing, China: China Statistics Press.

Han, J. (2009). Evolution of the Relationship between Rural and Urban Areas in the Past Sixty Years: Looking Back and Prospect. *Reform*, 11:5–14. In Chinese.

Ho, S. P., & Lin, G. C. (2003). Emerging Land Markets in Rural and Urban China: Policies and Practices. *China Quarterly*, 175:681–707.

Jiangyin Statistics Bureau. (2004, 2010, 2016). *Jiangyin Statistics Yearbook*. Beijing, China: China Statistics Press.

Knight, J., Song, L., & Shi, L. (2006). The Rural-Urban Divide and the Evolution of Political Economy in China. *Marine Policy*, 14(3):210–213.

Kunshan Statistics Bureau. (2004, 2010, 2016). *Kunshan Statistics Yearbook*. Beijing, China: China Statistics Press.

Literature Research Office of the CPC Central Committee. (1984). *Building Socialism with Chinese Characteristics*. Beijing: People's Publishing House. (In Chinese).

Ma, J. X. (2008). *Urban and Rural Relationships: From Dual Structure to Integration*. Beijing: Party School of the Central Committee of CPC. (In Chinese).

Nanhai Statistics Bureau. (2016). *Nanhai Statistics Yearbook*. Beijing, China: China Statistics Press.

China Statistics Bureau. (2015). *Statistics Yearbook of Beijing, Tianjin and Hebei Province*. Beijing, China: China Statistics Press.

Tian, L., & Liang, Y. L. (2013). The Industrialization and Land Use in Peri-Urban Areas: An Analysis Based on the Development of Three Top 100 County Economies in Three Regions. *Urban Planning Forum*, 34(5):30–37. (In Chinese).

China Statistics Bureau. (2015). *Urban Agglomeration Statistics Yearbook of PRD*. Beijing, China: China Statistics Press.

Wan, S. W. (2008). Investigation and Thinking of the History of Contemporary Chinese Urban and Rural Relationship. *Journal of Guizhou Normal University (Social Sciences)*, 4:14–20. (In Chinese).

Wang, D. R., & Ren, Z. D. (2011). The Path Dependence and Path Lock-In: Comparison on the Changes in the Space System in Relation to the Socio-Economic Development in Pearl River Delta and Yangtze River Delta. *Journal of Urban and Regional Planning*, 4(1):69–78. (In Chinese).

Whyte, M. K. (2010). *One Country, Two Societies, Rural-Urban Inequality in Contemporary China*. Cambridge, MA: Harvard University Press.

Xia, Y. X. (2008). The Evolution and Thinking of Urban-Rural Relations in China in the Past 30 Years of Reform and Opening-Up. *Journal of Suzhou University (Philosophy and Social Sciences)*, 11:18–20. (In Chinese).

Zhang, G. W. (2001). The Historical Investigation and Characteristics Analysis of Chinese Dual Structure's Evolution. *Macroeconomics*, 8:33–38. (In Chinese).

4 Industrialization, fragmented peri-urbanization, and land-use dynamics

This chapter reviews the rural industrialization process driven by township–village-enterprises (TVEs) and the burgeoning industrial clusters dominant in urban states after the demise of TVEs, and the various forms of city expansion in the context of the social-economic transition. Rapid urbanization in peri-urban areas has been driven by both the spill-over effects of the central city and non-agricultural land growth led by numerous autonomous rural collectives. This urbanization in peri-urban areas has resulted in a fragmented landscape. This chapter selects three representative peri-urban areas from three major urban agglomerations across China: Shunde in the PRD, Jiangyin in the YRD, and Shunyi in the BTH Region. By analyzing the land-use dynamics of typical peri-urban areas, this chapter identifies the spatio-temporal characteristics of land-use change such as land fragmentation, mix of residential and industrial land, and fragmented property rights. This chapter also examines the driving force behind land-use change in peri-urban areas.

4.1 Industrialization facilitated by TVEs and industrial clusters in peri-urban areas

4.1.1 Emergence of TVEs

In the early 1980s, the Household Production Responsibility System gradually replaced People's Communes in rural China and significantly improved farm productivity in rural areas (Putterman, 1993; Kung, 1995). Correspondingly, the problem of surplus rural labor emerged, providing opportunities for the rise of TVEs that were in need of an industrial workforce, thus beginning the first wave of rural industrialization (Zhu & Guo, 2015). Market-driven rural industries blossomed in some dynamic regions, providing unemployed farmers with industrial employment (Chang & Kwok, 1990). In 1987, rural industries exceeded agriculture in the contribution to total rural income at the national level (Oi, 1999). In 1994, the proportion of TVEs as part of China's total industrial production was 38%, more than quadrupling the 9% it had been in 1978 (Chang & Wang, 1994). Collective rural industries developed rapidly and soon became the backbone of the rural economy in peri-urban areas, as well as significantly improving the quality of life for residents in peri-urban areas. The rise of rural collective industries had been an important hallmark in the long history of agrarian China, indicating its transition from an agricultural civilization to industrialization (Zhu, 2017).

4.1.2 Decline of TVEs and emergence of industrial clusters in peri-urban areas

From 1985 to 1995, the proportion of village-initiated industries as part of total industrial output decreased while the share of township industries increased consistently in some peri-urban areas, especially in the YRD where a sense of clanship was not entrenched (Zhou, 1998). Townships, which were better at coordinating external industrial expertise and serviced sites, had taken a leading role at the beginning of rural industrialization in the YRD (Zhu, 2017). Moreover, with the rise of foreign enterprises and market integration, TVEs were no longer competitive. In the mid-1990s, China witnessed a significant decline of village industries and the emergence of industrial parks which were mainly managed by municipal and town governments. Between 1989 and 1996, most TVEs were dismantled and transformed into stock cooperative companies or private business, and this was mainly because of severe environmental problems both in rural and urban suburban areas caused by the scattered and small-scale plants and the financial burden of collectives' bad debt (Saich, 2001). At the national level, TVE reform was first initiated in Shunde County in the PRD, and then spread to the YRD and other region of China. In the late 1990s, the main industrialization tendency was shifted to inward foreign direct investments and joint-ventures in peri-urban areas.

Village industries were scattered, while foreign and private investments were located in planned industrial zones arranged by governments with preferential policies (Wei & Zhao, 2009). The foreign and private investments brought increasingly fierce competition to rural-initiated industries which were mostly rudimentary and immature. Thus, concentrated industrial zones managed by cities or townships gradually replaced scattered village factories as the main feature on the rural industrial landscape. Favorable policies and efficient infrastructure helped strengthen industrial zones and promoted further industrial agglomeration, while village-initiated industries gradually lost their competitiveness. As a result, enterprises in most village collectives were privatized or went bankrupt in the late 1990s. This led to a significant decrease in the number of village-initiated enterprises and their contribution to total rural industrial production, indicating the shift of peri-urban industrialization from village-initiated to municipality coordinated.

Most village-initiated industries in peri-urban areas were closely connected with local government and some have adopted new forms of "industrial clusters" (Naughton, 2007; Huang, 2008). The development of village-initiated industrialization in peri-urban China heavily relied on manufacturing, while technology-oriented industries lacked a suitable environment for developing. In many years, manufacturing often accounts for 60–70% of gross domestic product (GDP), and sometimes even higher (Tian, 2015). Although village-initiated industry is dynamic, it suffers from inherent drawbacks. On the one hand, subject to the uncertainty of property rights over collective land, village-initiated industries tend to pursue short-term benefits more than long-term sustainable development, and are unwilling to invest in industrial upgrading and regional joint developments. On the other hand, with high population density and small-scale village autonomy, village-initiated industries develop independently lacking comprehensive planning, resulting in fragmented industrial sites and urbanization which generate problems of industrial sprawl and compromised industrial productivity (Zhu, 2017).

Meanwhile, with the rapid rise of land value in city centers, manufacturing factories which used to locate in the city proper move to industrial parks in urban peripheral areas in order to get larger space and low-cost land. Similarly, some public facilities such as universities and hospitals find it difficult to expand their scale, and they have had to move to peri-urban areas, accompanied with city growth. Moreover, the establishment of national, provincial, or municipal level economic and technological development zones (ETDZs) outside large cities drives land development in peri-urban areas (Webster & Muller, 2002), helping to boost the economy of peri-urban areas and creating numerous job opportunities (Tian, 2015). Therefore, the combination of bottom-up rural industrialization and city spill-over effects jointly contributes to the development of peri-urban areas.

4.2 Peri-urbanization characterized by fragmented land use and governance

4.2.1 Fragmented land use in peri-urban areas

Due to historic reasons, rural settlement has been spatially scattered in peri-urban areas, and industrial land has been fragmented because of the legacy of dispersed TVEs development. Therefore, the intermingled landscape of residential and industrial land has been prevalent (Tian et al., 2017), which is called as "family workshop, plants village" (*jiajiadianhuo, cuncunmaoyan*). Moreover, due to good location and low land cost, peri-urban areas have been the destination of many industrial parks built by municipal or township government. As a consequence, state-led top-down industrialization and bottom-up rural industrialization coexist, which has led to the increasingly mixed degree of residential and industrial land (Tian, 2015). While the mix of residential and commercial, office and recreational activities is desirable for reducing commuting distance, the mixed residential and industrial land has generated many negative environmental consequences.

Wang et al. (2014) pointed out that pollution caused by manufacturing in rural areas has been one of big challenges China has been facing. From 1988 to 2010, the pollution industry took a share of 63.15% in all industry types and generated a large amount of waste water, waste gas, and solid waste without appropriate disposal. Li (2017) confirmed that industrial pollution is serious in rural areas. In 2008, around 274,500 polluting enterprises were distributed in approximately 227,000 villages, and there were around 387,000 manufacturing enterprises in peri-urban areas. Most of them were spatially scattered and small-scale factories, and were not equipped with necessary environment protection facilities.

4.2.2 Fragmented governance in peri-urban regions

Peri-urban regions have been characterized by a mix of state-owned and collectively owned land, and its governance involved multiple layers. For instance, most state land is managed by city governments, and collectively owned land has been managed by township, administrative village, and natural village. Motivated by the desire to push forward economic growth, each level of government, from city/county to village cadre, has been committed to industrial development (Figure 4.1). Under the urban–rural dual land management system, local revenues in peri-urban areas heavily relied on rent from land leasing, resulting in intensive land rent capture. Since the 1990s, a large amount of agricultural land has been converted into industrial use in a spatially fragmented and scattered pattern.

The establishment of economic and development zones and industrial parks has been a major tool of economic development in China. Driven by the incentive of maximizing tax revenue, the city/county government has been attracting outside investment to locate in their industrial zones. These industrial parks usually cover a large land area, ranging from several square kilometers to tens, even hundreds of square kilometers, and only those enterprises which can meet minimum investment and required environmental standards are allowed to enter the industrial park.

Meanwhile, townships are interested in building their own industrial clusters. Given egalitarian principles, each township government is usually allowed to establish one or two industrial clusters. According to Tian (2015), the land in township industrial clusters is usually mixed state-owned and collectively owned. Those enterprises with economic strength consider that state-owned land is more secure than collectively owned land, and they prefer paying land conveyance fee to convert collective land into state land. In general, the higher the governmental authority, the more stringent the planning control. As a result, the land use is more concentrated.

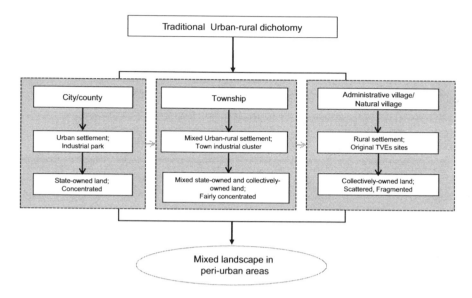

Figure 4.1 Governance layers and land-use characteristics in peri-urban areas.
Source: Tian, L. (2015)

4.3 Spatio-temporal characteristics of land-use dynamics in three peri-urban areas

4.3.1 Study area and data source

Given the data availability, this chapter selects three peri-urban areas, Shunde in the PRD, Jiangyin in the YRD, and Shunyi in the BTH region as case studies to compare their land-use change characteristics. The data of land use for Shunyi is from the Beijing Urban and Rural Planning committee, and is based on plot-based investigation. The data of land-use for Shunde and Jiangyin is from land-use information based on the topographic map at a 1:10,000 scale and site survey in 2001 and 2010 when their city master plans were made and later revised. Compared with Landsat thematic mapper (TM) satellite images which cannot differentiate various types of construction land such as residential, commercial, and industrial land, the land-use map can provide detailed information of land-use types, and its resolution is much higher.

4.3.2 Land-use structure changes in peri-urban areas

Based on the land-use investigation, non-agricultural lands in peri-urban regions are categorized as residential land, industrial land, commercial land,

and land for public facilities. Land for public facilities comprises land for public services and recreation, but doesn't include land for infrastructure.

Table 4.1 reveals the changes of land-use composition in the three peri-urban areas over the span of ten years, namely, from 2000 to 2010 for Shunde and Jiangyin and from 2004 to 2014 for Shunyi. Figure 4.2 presents the changes of land-use structure. On the whole, non-agricultural land increased significantly. The proportion of non-agricultural land of total land increased by 1.55, 4.75, and 2.72 times in Shunde, Jiangyin, and Shunyi, respectively. The growth of non-agricultural land in Jiangyin was highest among the three areas.

In terms of land-use structure, the proportion of residential land to total land was 15.70% in Shunde, 16.33% in Jiangyin in 2010, and 16.8% in Shunyi in 2014. In Jiangyin and Shunyi, the proportion of residential land increased rapidly by 13.46% and 15.37%, respectively, over ten years, while the growth of residential land was much slower in Shunde, at a growth rate of 4.29% over ten years. This is because development in the PRD jumped ahead of that of the YRD and the BTH region. The proportion of industrial land in total land reached 13.74% and 12.99% in Shunde and Jiangyin in 2010, and the growth of industrial land in Jiangyin was higher than that in Shunde during the ten-year period. However, the percentage of industrial land of total land remained below 5% in Shunyi from 2004 to 2014. On the one hand, the percentage of non-agricultural land was much lower in Shunyi than in Shunde and Jiangyin; but on the other hand, compared with Shunde and Jiangyin whose economic structures were dominated by the

Table 4.1 Land-use structure in three peri-urban areas

	Shunde		Jiangying		Shunyi	
Year	2001	2010	2001	2010	2004	2014
Residential Land	92.16 (11.43%*)	126.08 (15.64%)	28.41 (2.88%)	160.09 (16.21%)	14.57 (1.44%)	171.54 (16.8%)
Industrial Land	49.02 (6.08%)	110.29 (13.68%)	26.38 (2.67%)	127.30 (12.89%)	36.53 (3.61%)	43.91 (4.3%)
Commercial Land	13.99 (1.74%)	12.96 (1.61%)	5.35 (0.54%)	14.82 (1.5%)	4.76 (0.47%)	4.84 (0.47%)
Public Facilities Land	44.59 (5.53%)	58.60 (7.27%)	7.46 (0.76%)	16.91 (1.71%)	13.63 (1.35%)	15.38 (2.14%)
Others	606.24 (75.22%)	498.07 (61.79%)	919.26 (93.16%)	667.73 (67.68%)	950.40 (93.19%)	784.21 (76.89%)
Total	806		986.86		1,011.5	

* *means the percentage in total land*

Unit: *Sqr.km*

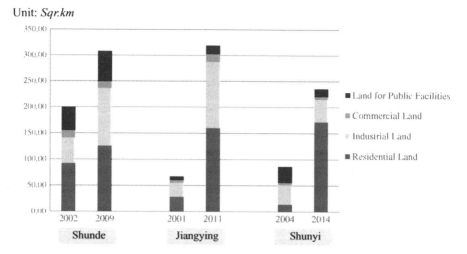

Figure 4.2 Land-use structure changes in three peri-urban areas
Source: drawn by authors

manufacturing industry, the service industry took a larger share in the economic structure of Shunyi due to the service-economy affiliated with Beijing
Capital International Airport located in Shunyi.

Commercial land and land for public facilities accounted for a small share,
less than 2% in land-use structure for the three peri-urban areas. Compared with
Shunde and Jiangyin, the share of commercial land in Shunyi was much lower,
less than 1/3 in Shunde and Jiangyin. Meanwhile, the growth of commercial
land was less than 1% in the three regions over a decade's period, among which
the growth of Jiangyin was highest. Land for public facilities grew by around
1%, much lower than that of residential land in the three peri-urban areas.

In summary, non-agricultural land increased significantly in the three
areas and was highest in Jiangyin. Among the four types of non-
agricultural land, the growth of industrial land was dominant in Shunde,
and the growth of residential land was highest in Shunyi, while in Jiangyin,
both residential and industrial land grew significantly. Compared with
Shunyi, the characteristics of peri-urban areas are much stronger in Shunde
and Jiangyin due to bottom-up rural industrialization. Land use in Shunyi
is affected most by the growth in its real estate industry.

4.3.3 Land-use pattern changes in peri-urban areas

In order to better understand the characteristics of land-use pattern change
in peri-urban areas, we adopt landscape ecology indices to quantify the
characteristics of land-use patterns, and they are explained as follows:

- Percentage of landscape (PLAND): the percentage of land area of land-use type i, Ai, in total land area A. PLAND = Ai/A.
- Number of patches (NP): number of patches of land use i. The larger the NP is, the more fragmented the landscape.
- Mean patch size (MPS): overall area of patches of land use i divided by number of patches of land use i. MPS = Ai/ni. The smaller the MPS, the more fragmented the landscape.
- Patch density (PD): total number of patches of land-use i divided by the total landscape area A. PD = ni/A. PD is also used to measure the fragmentation level of land use. The larger the PD, the more fragmented the landscape.
- Landscape shape index (LSI) is used to measure the regularity of the landscape shape. LSI = $(0.25/\sqrt{S}) \times L$, where L is the total length of edge (or perimeter) of land use i, and S is the area of class i. The closer to 1, the more regular the shape.
- Landscape fragmentation degree (LFD): number of patches of land use i divided by overall area of patches of land use Ai, LFD = $(ni- 1)/Ai$. The bigger the LFD, the more fragmented the landscape.

Table 4.2 presents the characteristics of four categories of land use in Jiangyin and Shunde in 2001 and 2010 and that of Shunyi in 2004 and 2014. There are several findings as follows:

For the total of non-agricultural land, the PLAND in Shunde, Jiangyin, and Shunyi all increased over the span of ten years. Among them, Jiangyin increased by 20%, while the growth of Shunde was relatively small. In terms of land-use indices, the MPS of non-agricultural land in Shunde increased, NP, PD, and LFD declined, and LSI showed a trend of approaching 1. In general, land fragmentation was reduced in Shunde from 2001 to 2010 and there was a tendency of aggregation. In Jiangyin and Shunyi, the NP and PD of non-agricultural land both increased over ten years, indicating an increasing trend of non-agricultural land patches. The MPS and LSI in Jiangyin remained basically unchanged, while MPS declined and patch shapes became more complicated in Shunyi. The LFD in Jiangyin and Shunyi have both increased. On the whole, there is an increasing fragmentation degree in the total non-agricultural land in both Jiangyin and Shunyi.

For residential land, PLAND and MPS increased while NP, PD, and LFD declined in Shunde from 2001 to 2010. This indicates an aggregating trend in residential land use in Shunde. In Jiangyin, there was a dramatic rise in PLAND, NP, PD, and LFD during the ten-year period, with MPS and LSI remaining unchanged. In other words, new residential land in Jiangyin has not expanded on the basis of existing land-use patches, but increased in the form of new separate patches, leading to fragmented land use. Similarly, Shunyi displayed a trend of fragmentation from 2004 to 2014, with more significant increases in PLAND, NP, PD, and LFD than in Jiangyin. The decline of MPS also confirms that the area of newly added

Table 4.2 Changes of land-use pattern in three peri-urban areas

Shunde

Year	2001					2010				
Land-use type	NAL	RL	IL	CL	PF	NAL	RL	IL	CL	PF
PLAND	24.75%	11.42%	6.07%	1.73%	5.52%	38.34%	15.70%	13.74%	1.61%	7.29%
NP	6,846	5,793	1,862	1,454	2,378	4,350	3,769	1,147	572	327
MPS (ha)	2.9	1.6	2.6	1.0	1.9	7.1	3.3	9.6	2.3	1.8
PD	8.48	7.18	2.31	1.80	2.95	5.42	4.69	1.43	0.71	0.41
LSI	0.86	0.98	0.92	0.89	2.04	0.97	0.91	1.13	1.09	0.97
LFD	2,345.8	3,640.9	706.9	1,509.6	1,267.6	614.1	1,126.4	119.2	252.0	181.9

Jiangyin

Year	2001					2010				
Land-use type	NAL	RL	IL	CL	PF	NAL	RL	IL	CL	PF
PLAND	7.16%	2.88%	2.67%	0.54%	0.76%	32.56%	16.33%	12.99%	1.51%	1.73%
NP	1,076	570	692	338	307	5,662	4,960	2,405	1,050	679
MPS (ha)	6.6	5.0	3.8	1.6	2.4	5.6	3.2	5.3	1.4	2.5

	2004					2014				
PD	1.09	0.58	0.70	0.34	0.31	5.78	5.06	2.45	1.07	0.69
LSI	1.13	1.22	1.01	1.11	0.91	1.21	1.15	0.98	1.04	0.84
LFD	163.7	114.1	181.3	213.1	125.9	1,004.4	1,536.4	454.2	743.1	272.2

Shunyi

Year	2004					2014				
Land-use type	NAL	RL	IL	CL	PF	NAL	RL	IL	CL	PF
PLAND	8.47%	1.43%	3.58%	0.47%	2.99%	23.71%	16.80%	4.30%	0.47%	2.14%
NP	460	52	263	84	234	2,278	2,339	643	169	589
MPS (ha)	18.8	28.0	13.9	5.7	13.0	10.6	7.3	6.8	2.9	3.7
PD	0.45	0.05	0.26	0.08	0.23	2.23	2.29	0.63	0.17	0.58
LSI	1.78	0.89	0.99	1.02	0.91	2.16	1.09	1.00	0.93	1.00
LFD	24.4	1.8	18.9	14.7	17.9	214.2	318.8	94.0	58.6	158.4

Note: NAL=Non-agricultural land
RL=Residential land
IL=Industrial land
CL=Commercial land
PF=Land for public facilities

residential patches is small. Compared with Jiangyin, residential land use in Shunyi displayed a trend toward more fragmentation.

In terms of industrial land, there was a significant trend of aggregation in Shunde. From 2001 to 2010, the PLAND and MPS increased while NP, PD, and LFD declined. The trend showed that existing patches of industrial land in Shunde were expanding and becoming less fragmented. However, there was a trend of fragmentation in industrial land use in both Jiangyin and Shunyi. In Jiangyin, PLAND, NP, MPS, PD, and LFD increased dramatically, indicating the dual trend of aggravation and fragmentation. On the one hand, existing industrial patches have been expanding. On the other hand, many new patches emerged, resulting in fragmentation. In Shunyi, the PLAND, NP, and PD increased slightly while MPS declined, and the industrial land-use pattern became more fragmented.

For commercial land, the PLAND and LSI remained almost unchanged, and MPS increased, while NP, PD, and LFD decreased in Shunde, and there was a clear trend of integration among the various land uses. In contrast, the PLAND, NP, PD, and LFD of commercial land increased significantly in Jiangyin while MPS and LSI remained unchanged. We find that a number of new commercial patches emerged in Jiangyin, leading to fragmented land use. The PLAND of commercial land in Shunyi remained almost constant during 2004–2014. As for the pattern of commercial patches, NP and PD increased while MPS decreased. Although the patches became more regular in shape, the degree of fragmentation of commercial land has increased in Shunyi on the whole.

In terms of land used for public facilities, there was a slight increase in PLAND, while at the same time, a significant decline in NP and PD was seen in Shunde over the ten-year period. The patches changed little in size and became more regular in shape, showing a trend toward integration. In Jiangyin, MPS and LSI did not change much, but the NP, PD, and LFD increased significantly, indicating a scattered and fragmented trend. Unlike that in the other two regions, the PLAND, NP, PD, and LFD of land for public facilities have all increased, and patches became more regular in shape as the LSI was closer to 1. In other words, there is some growth in land for public facilities in Shunyi, but existing land patches became smaller as MPS decreased, demonstrating a fragmented land-use pattern over the ten-year period studied.

In summary, growth in the four categories of land was insignificant in Shunde from 2001 to 2010, showing a trend of aggregation. In Jiangyin, land-use change presented characteristics of both aggregation and fragmentation. On the one hand, existing patches expanded; while on the other hand, new patches grew rapidly in scattered forms. As more patches emerged, the overall trend became more fragmented. Similarly, the land-use pattern became fragmented in Shunyi from 2004 to 2014.

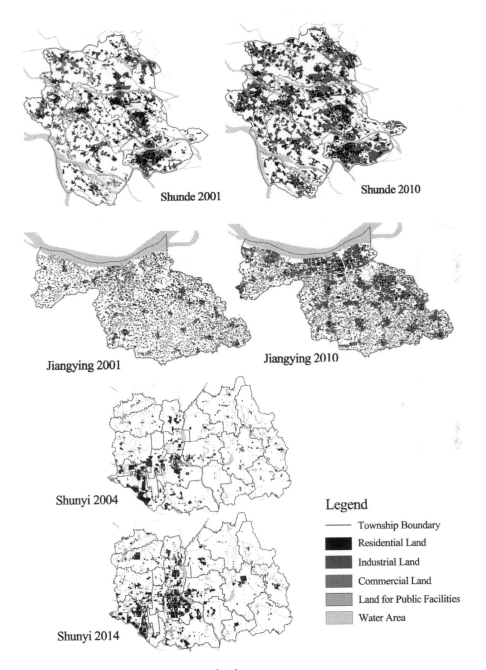

Shunde 2001

Shunde 2010

Jiangying 2001

Jiangying 2010

Shunyi 2004

Shunyi 2014

Legend

—— Township Boundary

Residential Land

Industrial Land

Commercial Land

Land for Public Facilities

Water Area

Figure 4.3 Land-use map in three peri-urban areas

Source: drawn by authors

4.3.4 Land-use diversity changes in peri-urban areas

The most frequently used index to measure diversity is Shannon's diversity index (SHDI) based on information theory (Shannon & Weaver, 1949), which is applied to quantify the degree of diversity at the landscape level. The value of SHDI represents the amount of "information" per patch. The absolute magnitude of SHDI is not particularly meaningful; therefore, it is used as a relative index for comparing different landscapes or the same landscape at different times (McGarigal et al., 2002).

$$\text{SHDI} = -\sum_{i=1}^{m}(P_i{}^{\circ}lnP_i)$$

Pi = proportion of landscape occupied by patch type (class) *i*. SHDI≥0, without limit; SHDI = 0 when the landscape contains only 1 patch (i.e., no diversity). SHDI increases as the number of different patch types (i.e., patch richness, PR) increases and/or the proportional distribution of area among patch types becomes more equitable.

According to Table 4.3, SHDI dropped from 1.22 to 0.95 in Shunde, from 1.18 to 1.10 in Jiangyin, and from 1.19 to 0.8 in Shunyi. These drops demonstrate that the diversity of landscape decreased in all three areas due to the emergence of new predominant land uses.

The SHDI decreased 0.17 between 2001 and 2010 in Jiangyin, and this was due to the fact that residential land and industrial land maintained a dominant position in both 2000 and 2010, and their proportion reached at 50% and 40%, respectively, in 2010. The proportion of commercial land and public facilities land fell to 5%, strengthening the dominant functions of residential land and industrial land, resulting in a reduced SHDI.

The SHDI decreased 0.39 between 2000 and 2010 in Shunyi, the largest decrease among the three peri-urban areas. The dominant land-use types in Shunyi were industrial land (42%) and public facilities land (35%) in 2004. By 2014, the proportion of residential land rose rapidly to 73%, becoming the dominant land-use type, and the equilibrium of the landscape was significantly reduced.

Table 4.3 SHDI of non-agricultural land in three areas

	Shunde		*Jiangyin*		*Shunyi*	
Year	*2001*	*2010*	*2001*	*2011*	*2004*	*2014*
SHDI	1.22	0.95	1.18	1.01	1.19	0.80

Table 4.4 reveals the proportion of non-agricultural land occupied by different patch types in each of the three areas. Residential land accounted for the highest proportion (46% in 2001 and 41% in 2010) among land types in Shunde. Meanwhile, industrial land took a share of 25% in 2000, rising rapidly to 36%, while the proportion of land for public facilities decreased from 22% in 2000 to 19% in 2010. In Jiangyin, the dominant land-use types were residential and industrial land, and their share increased from 81% in 2001 to 90% in 2010, and the proportion of industrial land reached 40% in 2010. Different from Shunde and Jiangyin, the proportion of residential land rose from 21% in 2004 to 73% in 2014 in Shunyi, while the proportion of industrial land dropped from 52.6% in 2004 to 19% in 2014. Shunyi has transformed from a typical peri-urban area into an urbanized area with extensive growth in the real estate industry. The investment in real estate rose from 3,630 million Yuan in 2004 to 24,550 million in 2014, growing by 676%. This was mainly attributed to rising housing prices and demand in the Beijing municipal area.

4.3.5 Residential and industrial land-use mix in peri-urban areas

In general, the mix of residential and industrial use can be classified into three situations: (1) residential and industrial use is mixed within family workshop or a home-based factory; (2) industrial zones are adjacent to residential areas; (3) factory/warehouse areas are intermingled with houses in peri-urban areas (Tian et al., 2017). Information on industrial parks and factory/warehouse areas can easily be captured by land-use maps, but the information on home-based factory or family workshop is not readily

Table 4.4 Land-use structure by land-use patches in three areas

	Shunde		Jiangyin		Shunyi	
Year	2001	2010	2001	2010	2004	2014
Residential land	92.16 (46%)	126.08 (41%)	28.41 (42%)	160.09 (50%)	14.57 (21%)	171.54 (73%)
Industrial land	49.02 (25%)	110.29 (36%)	26.38 (39%)	127.30 (40%)	36.53 (52.6%)	43.91 (19%)
Commercial land	13.99 (7%)	12.96 (4%)	5.35 (8%)	14.82 (5%)	4.76 (6.8%)	4.84 (2%)
Land for public facilities	44.59 (22%)	5.86 (19%)	7.46 (11%)	16.91 (5%)	13.63 (19.6%)	15.38 (7%)
Total	199.76 (100%)	255.2 (100%)	67.6 (100%	319.13 (100%)	69.49 (100%)	235.68 (100%)

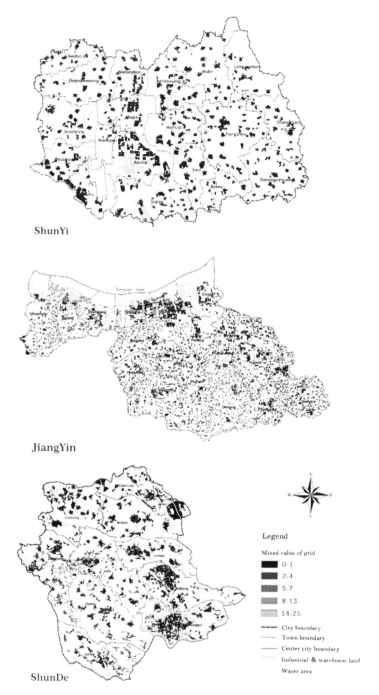

Figure 4.4 MDI of R&I land in three peri-urban areas

available. In this research, we mainly focus on the second and third types of land-use mix.

There has been a wealth of research measuring land-use mix and its bene-fits; however, there have been few studies on quantifying mixed degree of residential and industrial land (*hereafter* R&I). When planners are discuss-ing the negative externalities of this type of mix, they lack the indicator to precisely compare that land-use mix of R&I is more serious in this city than that city. In order to fill this gap, we developed an index method to measure the mix degree (*hereafter* MDI) of R&I land and obtained a patent granted by the State Intellectual Property Office of P. R. China (2016, Patent number: 201310659117.6). As for the detailed information how MDI is gen-erated, readers can refer to Tian et al. (2017).

The MDI of a particular region (*j*) can be calculated as follows:

MDI of a residential patch = actual assigned value of residential patch/ theoretical maximum assigned value (25). Mixed degrees of all grids are then totaled up so as to obtain the total MDI. It is expressed as follows:

$$\text{MDI}\,(j) \;=\; \sum_{i=1 \to n} \frac{n_i}{(n \times 25)} = \frac{1}{n \times 25} \sum_{i=1}^{n} n_i$$

where n is the number of grids covered by residential patches in a region j; n_i is the mixed degree of a residential patch i in the region j; 25 is the maximum possible assigned value for a residential grid cell.

Figure 4.4 presents the MDI of residential and industrial land based on land-use maps of Shunde and Jiangyin in 2010, and Shunyi in 2014. Table 4.5 shows the statistical result of MDI calculation in these three areas. There are totally 13,838 residential grids in Shunde, 13,001 in Jiangyin, and 12,000 in Shunyi. Based on grid value, we classify the mix degree into five levels: (1) zero mix given a value "1" (including the residential grid sur-rounded by residential, commercial land, etc. other than industrial land); (2) slight mix given a value of "2–4"; (3) medium mix given a value of "5–7"; (4) high mix given a value of "8–14"; (5) very high mix given a value of "15–25." Slight mix does not generate significant influence on living envir-onment, but areas with a medium mix and above can adversely affect the

Table 4.5 MDI of R&I land in three peri-urban areas

	R&I Mix in Urban Area	R&I Mix in Rural Area
Shunde	0.063517	0.071045
Jiangyin	0.071255	0.066572
Shunyi	0.052269	0.066138

living environment. The results reveal that the mixed degree of urban R&I use was highest in Jiangyin, and that of rural residential and industrial land was highest in Shunde.

4.4 What drives land-use dynamics of peri-urban areas in three peri-urban areas?

Peri-urban areas share many overall similarities in terms of fragmented land use and mixed residential/industrial land use, while also showing regional differences. Industrialization and urbanization are usually considered as two key factors of land-use change, and different level of governments and individual behavior of land users shape the land-use change (Long et al., 2008). In the three peri-urban areas, the combination of bottom-up and top-down industrialization has been the major driver of land-use changes. Moreover, local institutional arrangements and policies also exert dramatic influences on land use.

4.4.1 Impacts of regional development trajectories on peri-urban areas

Due to the vast size of China, the industrialization and urbanization pattern varies by region to a large extent, affecting the development of peri-urban areas across China's different regions. The PRD was the first to launch the reform opening, followed by the YRD, while the BTH region was the latest to launch reforms. Since the reform opening, three pilot economic zones were established as experiments in Shenzhen, Zhuhai, and Shantou of Guangdong Province in 1980, and the PRD Economic Development Zone was set up in 1985. Afterwards, Guangzhou was established as an experimental area for economic system reform in 1988. Since that time, the economy witnessed great strength in the PRD. In the YRD, economic growth began with the opening-up of the Pudong New Area, Shanghai in the early 1990s. At that time, national priority shifted from the PRD to the YRD, facilitated by the "Pudong's Development and Opening-up" initiative and permission to build a capital market in 1992. The power engine of the BTH region was ignited by the establishment of Tianjin Binhai New Area in 2006. After 2006, the BTH region was shown favorable policies by the national government and endowed with more autonomy.

The three urban agglomerations studied have adopted different industrialization models. Economic growth of the PRD is mainly driven by export-driven exogenous forces, and it has been a major manufacturing base for products, in particular, electronic products. In 2001, nearly 5% of the world's goods were produced in the Greater PRD with a total export value of US$289 billion, and more than 70,000 Hong Kong companies built factory plants in the PRD.[1] Meanwhile, the YRD model is typically foreign investment oriented, and the share of Foreign Direct Investment (FDI) in the

YRD reached 47.87% of all foreign investments in China (*Source*: China Statistics Yearbook, 1980–2017). In the BTH region, economic growth is driven more by the investment of governments and SOEs, which is more domestically oriented.

Among the urbanization modes in different regions across China, the YRD began with urbanization of rural areas. This pattern highlights the development of small towns and relies on the development of non-agricultural business in rural areas. With the active and in-depth engagement of grassroots governments, this pattern combines the "top-down" and "bottom-up" urbanization modes (Wen, 2011). The urbanization of the PRD presents an example of the typical bottom-up model in which local governments, township enterprises, and surplus laborers are key drivers for urbanization. In the BTH region, urbanization is driven by top-down forces, and investments from central and local governments play an essential role in economic development, whereas township enterprises have limited influence on urbanization.

In summary, the PRD is mostly dominated by private enterprises and small and medium enterprises, while economic development in the YRD is built upon SOEs and foreign-funded companies, supplemented by support from the private sector. In contrast, the economic structure of the BTH region sees large and medium-sized SOEs account for a far higher proportion than the national average level, whereas private sector is growing slowly. The YRD boasts a full range of industrial categories with highly developed heavy and light industries. The BTH region gives a high priority to heavy industries, represented by metal processing and ore processing. In the PRD, the light industry is quite developed (Xie et al., 2008).

Moreover, the extent of government intervention has varied among the three urban agglomerations. From the north to the south, these three areas represent different stages of the transition from a planned economy to a market economy. The BTH Area is imprinted with apparent marks of the planned economy, as local governments and officials are still accustomed to directly intervening in the economic development using administrative means. By contrast, the PRD features many aspects of a market economy, as local governments and officials are rarely found to directly intervene in the market. Clues of both a planned economy and a market economy can be found in the YRD, as manifested by the coexistence of administrative intervention and market regulation (Liu et al., 2011). In summary, the PRD adopts a "decentralized governmental administration model." The YRD adopts a "government-led market model" characterized by a strong government and a strong market. The BTH region is an opposite of the PRD, as its government is playing a larger role than the free-market in economic development, which is characterized by a strong government and a weak market (Figure 4.5).

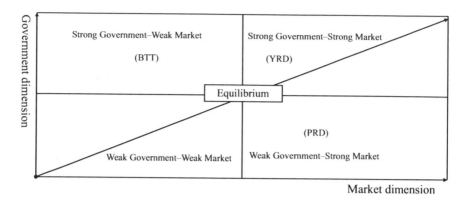

Figure 4.5 A diagram of government–market relationship
Source: drawn by authors

4.4.2 *Land financing of peri-urban areas under the tax-sharing system*

Since the reform opening, the TVEs in the rural area had played an essential role in boosting the Chinese economy, and its proportion in national industrial output rose from 11.11% in 1980 to 30.82% in 1991 (*Source*: China Statistics Yearbook, 1980–2017). During the period of 1978–1993, China adopted the fiscal contracting system, which granted local governments residual rights over local revenue (Oi, 1999). TVEs were a major income source for township governments and village collectives. In 1994, a new tax-sharing fiscal scheme was introduced to replace the fiscal contracting system, and tax revenue was shared among the central government, the provincial government, and the city/county government, and the central government took the largest share. In 2011, the central government took 54.2% of overall tax revenue, and the remaining 45.8% was shared among the provincial government and the city/county and township government (*Source*: China Tax Statistics Yearbook, 1980–2017).

Generally speaking, the share of tax revenue varies from city to city, from township to township, and the specifics of the tax responsibility system are too diverse and complex to detail. For example, in the Beijiao township of Shunde, the district government retained 25.5% of tax revenue, the township retained 6.5% of tax revenue in 2010, and the other 68% went to the higher levels of government (*Source*: Interview with local township government). In the Xinqiao township of Jiangyin city, the township retained 15.9% of tax revenue in 2012. In Shunyi District, the district government can only retain 12.5% of tax revenue, and less than 5% of tax revenue goes to village collectives (*Source*: Interview with district government). Since the majority of tax goes to the pocket of higher level government, the township government and village cadres are inclined to look for other income sources, and rent from land and property has become such a tool. Under this situation, township leaders and village cadres are willing to lease out their land, or build factory and commercial buildings to earn rent.

According to the statistics of village collective income in the three peri-urban areas, we find that land rent and property rent took the largest share at 65% in the collective income of Jiangyin in 2008, and other income derived from management fees of local enterprises or tenant enterprises. In 2012, this share rose to 65.92%. In Shunde, 69.42% of collective income came from land and property rents in 2008, slightly higher than that of Jiangyin. Management fees for local or tenant enterprises accounted for 19.01%, and the remaining 11.57% came from the income of running markets. In 2012, the share of land and property rent dropped slightly to 68.03% (Tian, 2015). In the Beishicao township of Shunyi, 58% of village income was from land and property leasing in 2009 (*Source*: calculated based on the data of Shunyi Agricultural Committee).

In China, the LURs conveyance fee has become an increasingly important source of local revenue. From 2000 to 2016, the proportion of LURs conveyance fee in total revenue accounted for 32.42% (*Source*: China Statistics Yearbook, 1980–2017). Under the stringent land quota system, more land quota has been allocated for industry when the local revenue base relies more on VAT and less on business taxes (Cai, 2012). Therefore, the expansion of industrial land has been prevalent in Shunde and Jiangyin. Comparing land-use change of three typical peri-urban areas over the ten years' span, we find that they share similar characteristics of land fragmentation induced by top-down and bottom-up industrialization, and fragmented governance. Meanwhile, they have shown some differences: land-use change in Shunde has shown a more consolidated trend than that in Jiangyin. This is in part due to the influence of land consolidation policies which have shifted from Greenfield development to the combination of redevelopment and Greenfield development. Shunyi is different from Shunde and Jiangyin since its real estate industry significantly developed due to the spill-over effect of the Beijing central city.

4.4.3 Government's preference for industrial land use under the land quota system

In China, GDP growth has been a key index to evaluate the performance of local officials. As a major source for generating GDP and employment, the manufacturing industry has been always the focus of local economic growth (Cao et al., 2008). Under the tax-sharing fiscal scheme, 60% to 70% of a local government's tax base is composed of value-added tax (VAT), business tax, and income tax from local enterprises other than central state-owned enterprises (Cai, 2012). Meanwhile, the VAT contribution of manufacturing industry accounted for around 60% of all VAT from 2001 to 2006 (*Source*: China Tax Yearbooks 2002–2007). Compared with the land leasing revenue, the VAT has been a sustainable, instead of once-for-all income source of local revenue.

Therefore, both city/county governments and township and village cadres are committed to attracting investments in the manufacturing sector.

Meanwhile, in 2003, the central government realized the significance of land policy in macro-social and economic development, and declared that they would apply land policies as a key part of national macro-control measures in order to protect farmland and improve industry structure through forbidding land from leasing to projects which are not consistent with national industry policy, development planning, and entry standard (Tian & Ma, 2009). At the same time, the central government strictly enforced a land quota system that restricts the maximum amount of newly added land converted from agricultural use to non-agricultural use. At the subnational level, land quotas are allocated, but sectoral allocation of land quotas is within their jurisdictions. Incentivized by generating more GDP and employment, local government has been inclined to allocate more land quota to industry where the local revenue base relies more on VAT and less on business taxes (Cai, 2012). For instance, Jiangyin no longer reserved land quota for new *Zhaijidi* in rural areas in order to leave enough quotas for industrial land. From 2001 to 2010, industrial land grew by 4.57 times, and rural settlement only increased by 1.12 times. During the same period, industrial land increased by 2.05 times in Shunde, while rural settlement rose by 1.58 times. Industrial land increased by 20.2% in Shunyi from 2004 to 2014, while residential land increased by 11.77 times. This is mainly because with the increasing housing demand price of Beijing, many residents have to move to suburban areas for cheaper housing.[2]

4.4.4 Bottom-up institutions, local policies, and their impacts on land use

Compared with Shunyi where the impact of spill-over effects of the central city are more apparent, Jiangyin and Shunde share more similarities in local governance, rural industrialization, and land fragmentation. However, due to the divergence in land policies, land-use change in Jiangyin and Shunde has shown different characteristics. Over the ten years' span, rural settlement and industrial land in Shunde have displayed a much more consolidated trend than those of Jiangyin and Shunyi, which is mainly attributed to different land policies adopted in these areas.

(1)   Land consolidation policies in Shunde

Shunde launched their land consolidation program in 2000, and designated 17 industrial clusters managed by the county and township. The government required new industrial developments to be located within these 17 industrial clusters (Tian, 2015). Meanwhile, the Shunde government issued a series of policies to close the village industrial areas. For instance, it set up a fund to encourage the relocation of enterprises in village industrial areas to industrial clusters. In 2007, Shunde initiated their "Three-olds Renewal" Programs which were designed to facilitate the redevelopment process of old factories, towns, and villages in order to improve land-use efficiency, and supporting policies such as tax breaks and subsidies were also made to speed up the redevelopment process. The data of declining NP of rural settlement (from 1524 to 800) and industrial land (from 852 to 791) from 2001 to 2010 reveal that the shift from Greenfield development to redevelopment has been on the agenda of the Shunde government.

In 2001, Shunde released a policy on rural *Zhaijidi*, titled "Fixed *Zhaijidi* for rural household." This policy stipulated that the maximum *Zhaijidi* area for each qualified household member would be 80 m², and by the end of 2008, newborn children would no longer be allocated *Zhaijidi* in order to control the sprawl of rural settlements. As a consequence of these policies, industrial land increased by 2.06 times in Shunde, much lower than that in Jiangyin, and rural settlement rose by 1.58 times, slightly higher than that in Jiangyin.

Riding the wave of China's reform and opening-up, Shunde proposed to build city-level and town-level industrial zones, and launched policies of "Closing down and merging with others" (*Guantingbingzhuan*) for village industrial areas in 2000. In order to encourage industrial projects to settle in the industrial parks, Shunde introduced a series of preferential policies in 2002 for the purpose of attracting investment into township-level industrial parks. In 2007, Shunde called for redevelopment of old factories in village industrial areas, and provided special supporting funds for "Three-olds Renewal." For instance, enterprises or individuals could be rewarded with over one million Yuan for relocating industries (Tian, 2015).

Since the 1980s, with an increasing demand for land generated by rural industrialization in the PRD, village collectives found it was necessary to accumulate the land separately hold in individual farmers under the Household Contract Responsibility System (HCRS). Land Share-holding Cooperatives (LSCs) emerged in the early 1990s in order to assemble land from individual households who are allocated stakes based on their land contribution. Then, village collectives made the land-use plans, designated boundaries of farmland, commercial, industrial land, etc. Moreover, the LSCs built roads and infrastructure facilities such as water pipes, sewer pipes, gas, and other public facilities, and the cooperatives then leased the land out to outside enterprises. They managed land asset investment, collected land rents and other management fees, and distributed dividends to shareholders (Tian & Zhu, 2013). As a consequence, industrial land has been relatively concentrated at the village level in Shunde. With these land consolidation programs and bottom-up institutional arrangements, Shunde has achieved the goal of less fragmented land use than that found in Jiangyin and Shunyi.

(2) Land policies in Jiangyin

As early as 1985, Shanghai was aware of problems of disorderly and scattered land use and put forward the "Three Concentrations" policy to promote integrated and compact land use, namely, farming concentrated in agricultural parks, rural *Zhaijidi* concentrated in residential areas, and factories concentrated within industrial clusters. This was done in the hope that sprawl of non-agricultural land could be curbed and scattered construction land could be converted into farmland. This policy was then introduced to Jiangsu Province. Since the beginning of the 21st century, governments at all levels of Jiangsu Province made their objectives of the "Three Concentrations" policy based on quantified indicators.

In Jiangyin, the "Three Concentrations" policy was officially released in 2001. Since that time, any newly added industrial land must be located within township industrial clusters, and is not allowed to locate outside the designated clusters. However, consolidation turned out to be more difficult and the gain in cultivated land significantly less than originally thought (Ho & Lin, 2003). New industrial clusters were established, but the original TVE sites were reserved as one of the major income resources of villages. Compared with land use of Shunde in the PRD, land use in Jiangyin was much more fragmented. "Three Concentrations" were mainly made for newly added land rather than stock land, and Jiangjin did not make land polices targeting the redevelopment of stock land. Since relocating factories is a very expensive endeavor, factory owners have not been incentivized to relocate the industrial land.

(3)   Land policies in Shunyi

Shunyi mainly relies on its favorable location near the Beijing Capital Airport to attract a large number of domestic and foreign industrial enterprises, but its development of township enterprises lags behind Jiangyin and Shunde. Before 2007, in order to attract more investment, local townships offered low-cost industrial land for non-local enterprises. As a result, industrial land expanded rapidly, but the land-use efficiency was very low (Tian & Liang, 2013). In order to improve land-use efficiency, the Shunyi District Government issued "*Notice on the Opinions on Implementation of Bidding, Auction, Listing and Transfer of Industrial Land*" in 2007. This policy stipulates that industrial land should be transferred only by means of bidding, auction, and listing and it has helped curb the sprawl of industrial land, as witnessed by the growth of industrial land from 2004 to 2014 at only 7.38 km^2.

Different from Shunde and Jiangyin whose land-use pattern has been dominated by industrial land, Shunyi's land-use change was characterized by the dramatic increase of residential land that grew from 14.57 km^2 in 2004 to 171.54 km^2 in 2014. Xiu et al. (2013) named this phenomenon as "Suburbanization of Beijing Real Estate Industry." Many urban residents who work in the central city could not afford the high housing prices in the central city and moved to suburban areas, leading to a rapid expansion of residential land. However, the land for public facilities grew much slower than that of residential land, only increasing from 13.63 km^2 in 2004 to 15.38 km^2 in 2014, indicating that the quality of public services is low in peri-urban areas.

4.5 Summary

Over the past few years, the peri-urban areas of China have gained more attention as more people have flooded into them. However, urban planners and policymakers have been influenced by the traditional urban–rural dichotomy. Given the ambiguous status and interstitial location of peri-urban areas, planners and policymakers cannot fully understand the land-use dynamics and influencing mechanism of socio-economic factors on land-use change, and thus cannot identify suitable solutions for land management in peri-urban areas (Webster & Muller, 2002).

China is the second largest country in Asia. Because of its large size, it is haunted by a great unbalance in regional development. In the inland regions, most counties/county-level cities are still in the initial and middle stages of industrialization. The modernization of counties is a key solution to the urban–rural dual structure and unbalanced regional development across China, which is also the catalyst for China's continuous development. Based on the past three decades experience of reform and opening-up, China's peri-urban areas did get precious experience in the process of industrialization, urbanization, and modernization, which is of important significance for the county-level economic development – an integral part of China's national economy. However, the development of peri-urban areas is confronted with unprecedented challenges in the 21st century. Among these challenges are extensive economic growth supported by an inflow of massive resources which is hard to sustain, while restriction caused by the scarcity of land resource is quite intense; urbanization is far behind industrialization, rendering it powerless to drive industrial transformation in peri-urban areas; peri-urban areas also suffer from serious environmental problems and lack sufficient attractions for talents; and administration capabilities are not compatible with economic development while social conflicts and social troubles frequently occur.

As Webster and Muller (2002) pointed out, peri-urbanization processes vary dramatically across countries such as in East Asian countries, and it also varies within regions of individual countries. These peri-urban areas which have pioneered the reform and opening-up in China have been the first to encounter developmental limits and bottlenecks. A comparative study of various nations and regions can help us better understand land use in peri-urban areas, and help policymakers and planners make policies adaptable to the long-term sustainable development of peri-urban areas.

Notes

1 China's low costs are on the rise dallasnews.com 19/08/2008. Retrieved 08/02/2012.
2 Housing prices increased more than ten times from 2000 to 2010 in Beijing. Source: https://wenku.baidu.com/view/60097763fab069dc51220105.html; accessed on 05/15/2018.

References

Cai, M. (2012). Local Determinants of Economic Structure: Evidence from Land Quota Allocation in China. *Social Science Electronic Publishing*, 1(2):223–224.

Cao, G., Feng, C., & Tao, R. (2008). Local "Land Finance" in China's Urban Expansion: Challenges and Solutions. *China &World Economy*, 16(2):19–30.

Chang, C., & Wang, Y. (1994). The Nature of the Township-Village Enterprise. *Journal of Comparative Economics*, 19:434–452.

Chang, S. D., & Kwok, R. Y. W. (1990). The Urbanization of Rural China, In Kwok, R. Y. W., Parish, W. L., Yeh, A. G. O., & Xu, X. (Eds.) *Chinese Urban Reform: What Model Now?*Armonk, NY: M.E. Sharpe, 140–157.

China Statistics Bureau. (1980–2017. *China Statistics Yearbook*. Beijing, China: China Statistics Press.

Ho, S. P., & Lin, G. C. (2003). Emerging Land Markets in Rural and Urban China: Policies and Practices. *China Quarterly*, 175:681–707.

Huang, Y. (2008). *Capitalism with Chinese Characteristics*. New York: Cambridge University Press.

Kung, J. K. S. (1995). Equal Entitlement versus Tenure Security under a Regime of Collective Property Rights: Peasants' Preference for Institutions in Post-reform Chinese Agriculture. *Journal of Comparative Economics*, 21(1):82–111.

Li, Y. (2017). Industrialization and Urbanization in Peri-Urbanized Rural Areas. *Urban Development Studies*, 24(3):89–94. (In Chinese).

Liu, J., Liu, X., & Chen, Z. (2011). Historical Institutions Affect the Current Policies of Local Governments: A Comparative Analysis of the Labor Policies in the Pearl River Delta Region, the Yangtze River Delta Area and the Pan-bohai Sea Region. *Journal of Gansu Institute of Public Administration*, 4:107–115. (In Chinese).

Long, H., Wu, X., Wang, W., & Dong, G. (2008). Analysis of Urban–Rural Land-Use Change during 1995–2006 and Its Policy Dimensional Driving Forces in Chongqing, China. *Sensors*, 8(2):681–699.

McGarigal, K., Cushman, S. A., Neel, M. C., & Ene, E. (2002). *FRAGSTATS: Spatial Pattern Analysis Program for Categorical Maps. Computer Software Program*. Amherst, MA: University of Massachusetts.

Naughton, B. (2007). *The Chinese Economy: Transitions and Growth*. Cambridge, MA: MIT Press.

Oi, J. C. (1999). *Rural China Takes Off: Institutional Foundations of Economic Reform*. Berkeley, CA: University of California Press.

Putterman, L. G. (1993). *Continuity and Change in China's Rural Development: Collective and Reform Eras in Perspective*. New York: Oxford University Press.

Saich, T. (2001). *Governance and Politics of China*. Urbana, IL: University of Illinois Press.

Shannon, C. E., & Weaver, W. (1949). *TheMathematical Theory of Communication*. Urbana, IL: University of Illinois Press.

Tian, L. (2015). Land Use Dynamics Driven by Rural Industrialization and Land Finance in the Peri-urban Areas of China: The Examples of Jiangyin and Shunde. *Land Use Policy*, 45:117–127.

Tian, L., Ge, B. Q., & Li, Y. F. (2017). Impacts of State-Led and Bottom-Up Urbanization on Land Use Change in the Peri-Urban Areas of Shanghai: Planned Growth or Uncontrolled Sprawl?*Cities*, 60(B):476–486.

Tian, L., & Liang, Y. L. (2013). The Industrialization and Land Use in Peri-Urban Areas: An Analysis Based on the Development of Three Top 100 County Economies in Three Regions. *Urban Planning Forum*, 34(5):30–37.

Tian, L., & Ma, W. (2009). Government Intervention in City Development of China: A Tool of Land Supply. *Land Use Policy*, 26(3):599–609.

Tian, L., & Zhu, J. (2013). Clarification of Collective Land Rights and Its Impact on Nonagricultural Land Use in the Pearl River Delta of China: A Case of Shunde. *Cities*, 35:190–199.

Wang, Y., Liang, L., & Mei, Y. (2014). Study on the Temporal and Spatial Characteristics of Rural Industry Pollution. *Journal of Henan University*, 44(4):428–435. (In Chinese).

Webster, D., & Muller, L. (2002). *Challenges of Peri-Urbanization in the Lower Yangtze Region: The Case of the Hangzhou-Ningbo Corridor*. Working paper, Asia/Pacific Research Center, Stanford, CA:Stanford University.

Wei, Y., & Zhao, M. (2009). Urban Spill Over vs. Local Urban Sprawl: Entangling Land-Use Regulations in the Urban Growth of China's Megacities. *Land Use Policy*, 26(4):1031–1045.

Wen, T. J. (2011). *Understanding Southern Jiangsu*. Suzhou: Suzhou University Press. (In Chinese).

Xie, S. L., Li, Q., & Fang, J. F. (2008). A Comparative Analysis of Economic Development Strategies among Bohai Rim, YRD and PRD. *Economic Outlook the Bohai Sea*, 5:35–37. (In Chinese).

Xiu, D. P., Lv, P., & Yan, B. Y. (2013). Characteristics and Model of Real Estate Suburbanization: Based on Beijing Cases. *China Real Estate*, 4:24–30. (In Chinese).

Zhou, X. H. (1998). *Tradition and Transformation – The Social Psychology of Jiangsu and Zhejiang Villages and the Change since Modern Times*.Shanghai: SDX Joint Publishing Company. (In Chinese).

Zhu, J. M. (2017). Making Urbanization Compact and Equal: Integrating Rural Villages into Urban Communities in Kunshan, China. *Urban Studies*, 54(10):2268–2284.

Zhu, J. M.,& Guo, Y. (2015). Fragmented Peri-urbanisation Led by Autonomous Village Development under Informal Institution in High-Density Regions: The Case of Nanhai, China. *Urban Studies*, 51(6):1120–1145.

5 Land development under institutional uncertainty and land rent seeking in peri-urban areas

Urban China and rural China are two institutionally distinct domains, and rights over collectively owned land are ambiguously delineated. The dual land management system has created institutional uncertainty, and effective state governance is absent in peri-urban areas. This chapter first analyzes the nature of ambiguity of property rights over collectively owned land, followed by an analysis of formal or informal institutional change given the ambiguity of collective land. The characteristic of institutional uncertainty and how institutional settings affect modes of land development and land rent competition are then discussed. Finally, the results of land development under different methods of land rent distribution are analyzed and compared among cases in different regions.

5.1 The incremental change of land institutions driven by multiple forces

5.1.1 General characteristics of transitional institutions during China's gradual reform

The reform opening started in the late 1970s has been gradual, guided by pragmatism, and the underlying philosophy is well reflected in a succinct Chinese adage, "crossing the river by groping for stones (mozheshitouguohe)." As Zhu (2005, p. 1373) presented,

> pragmatism implanted in the reform process determines that the transformation is incremental and gradual, as opposed to the drastic shock therapy adopted by the former Soviet Union and some eastern European countries. Gradualism is meant as an instrument to legitimize rather than undermine the existing political system.

That is, as Lau et al (2000, p. 122) noted, to implement a reform without creating losers who are likely to hamper the reform. Gradual reform pushes forward China's transition.

From the perspective of institutional change, gradualism leads to dualism, the co-existence of new and old institutions (Wang, 1994; Zhu, 2005). The coordination mechanisms of top-down directives (plan as old institutions) and bottom-up initiatives (market as new institutions) are at work (Zhu, 1999). With new institutions coming in and old institutions going out, institutional change in transitional China takes on the characteristic of transition. As Zhu (2005, p. 1371) made clear,

> transitional institutions bridge the old and new institutions, which exist not by design, but as a consequence of compromise and competition between status quo interests and forward-looking forces. Tentative in nature, transitional institutions will be phased out as the economic reforms progress to an advanced phase.

Because of their short-term characteristic, transitional institutions may not provide sufficient order and certainty in the emerging market. In some cases where old institutions have ceased to exist and new institutions are yet to be created, the institutional transition may induce an institutional vacuum (Zhu, 2013a). As a result, uncertainty and disorderly competition may occur, which is typically so in emerging land markets (Zhu & Hu, 2009).

5.1.2 Land ownership change by law

(1)   The change of landownership: defining the "collective"

According to the 1954 Constitution of the People's Republic of China, "the state has the ultimate claim over land in China." Based on this fundamental principle, the dichotomy of collective ownership of rural land and state ownership of urban land was finally legitimized in the later "People's Commune Movement" in 1962. In the dual framework, rural collectives and urban governments, both as agents of the central state, managed collectively owned land and state-owned land, respectively. If cities needed expansion, the only way to convert rural land for urban uses was through state expropriation and then administrative allocation to urban agents (Ho & Lin, 2003). Urban agents could not directly acquire land from the rural sector.

Since 1978, the land system has undergone constant change. This has been criticized by some scholars as both blurring the "rights" boundary between dual ownerships and leading to the ambiguity of collective property rights *per se* (Ho, 2001; Lin & Ho, 2005). The pre-reform independence of urban governments and rural collectives in the governance over urban and rural land could not be sustained any longer. According to Article 2 of Land Management Law (LML) (1988, 1999), "by law, rural collectives have the rights to use and manage land, but no rights to transfer land for compensation uses. The state may expropriate rural land in the public interest." Meanwhile, Article 39 and 11 of the law stipulates "confirmation and the change of land ownership should be authorized by the government at or above county level." Under this arrangement, urban governments are designated as leaders of city regions that include both urban centers and the rural hinterland (Hsing, 2006). Moreover, it centralized the management of rural land to urban authorities (Brown, 1995; Lin & Ho, 2005). Rural collectives' rights over land were defined as such that the development rights should be granted by the urban state, and the alienation rights restricted to the situation where the other party in the transaction was the urban state (Zhu & Hu, 2009). Therefore, the state is the *de jure* "land owner," whereas rural collectives were the nominal owners.

Property rights of rural collectives are thought of as ambiguously defined in land law because of the unclear definition of "collective." In the commune system before China's reform, collectively owned land belonged to entities in three tiers which were composed of the commune, the brigade, and the team, with the team being the basic unit (Ho, 2001; Tang, 2009). After the reform, the three entities were changed to town/township, administrative village, and natural village in name, respectively. Meanwhile, collective land ownership was first redefined in the 1988 LML: "land already owned by the farmer's collective of natural village and township is administrated by their corresponding economic organizations; the remaining land is administrated by the economic organizations of administrative villages." This was called by Krug (2004) as assigning property rights based on the "usufructs" principle. However, according to Ho (2001), in the legal structure, the ownership of the successor of the team (the natural village) is no longer self-evident, especially when the administrative village is assigned as an autonomous unit in the governance of rural society. Because the amount of land to which each entity is entitled has never been clearly defined, collective landownership is ambiguous in some scholars' opinions (Ho, 2001; Zhu & Hu, 2009).

The situation of an ambiguous definition of "collectives" has been gradually changed since the late 2000s when the central government announced the experiments of clarification of collective landownership in some areas. In later years, the central government gradually popularized the experiment experience throughout the country requiring that all provinces must finish the ownership redefinition before 2012. The landownership definition associated with Guangdong's urban renewal, as will be introduced in Chapter 6, is a typical practice of the abovementioned requirements. In the process of landownership definition, the articles of the 1988 LML about collective land ownership are followed and the principle of "usufructs" is adopted. Natural villages or administrative villages are defined as the owner of the land under their control. Land boundaries are clarified and the certificates of landownership are issued to related owners.

(2)   Land rights assignments between the "collective" and governments

Chinese land is governed by a dual framework of collective landownership and state landownership. According to China's Constitution, in principle, urban land belongs to the state with the urban government as its agent, and rural land is collectively owned by the agrarian community. As a key component in the package of economic reform, the 1988 amendments (Article 10) to the constitution formally promulgate that the use rights of urban land can be separated from its ownership and leased to developers or users for a fixed period of time after a lump-sum rental payment. The invention of the public leasehold has explicitly made state-owned land an economic asset as in the nature of an investment instrument. Property rights are clearly defined over land supplied under such a leasehold (Tang, 1989) and an emerging urban land market has been evolving ever since (Zhu, 2005).

However, China's collective landownership is a unique institution, created based on the Marxist doctrine that land should be treated as a means of production, not an economic asset. Thus, land, both agricultural and non-agricultural, is owned by the rural villages on the condition that it can only be used by the collective owners themselves for economic production. When land is developed for non-agricultural uses like industries, the rural village must obtain permission from the urban government at the county level or above (Byrd & Lin, 1990; Brown, 1995). Collective landownership is thus defined as such that the collective has neither the right to change its form and substance by developing it, nor the right to derive income from land by letting it out for non-agricultural activities without approval from the urban government (Po, 2008). The right to convert collective land for non-agricultural uses is restricted to the situation where the other party in the transaction is the state.

In spite of the reforms of rural agricultural land towards circulation according to market principles, land conversion for non-agricultural uses has always been strictly controlled by law, whether the rules of the central government about strict protection of farmland are enforced or not by local governments (Lin & Ho, 2005). That is to say, collective land rights are still incomplete because collective landowners do not possess the autonomous rights to develop land for non-agricultural uses (Zhu & Hu, 2009). It is rational to infer that there is an irresistible impulse from the bottom-up to compete for land development rights with the government, which is evidenced by the existence of informal markets as shown in Sections 5.2.1 and 5.3.

(3)   Land rights assignments inside the collective

Individual peasants' rights over collective land can be discussed along two dimensions: discussing rights which are entitled to by the law, and discussing the relationships between peasants and village collective organizations with respect to land rights assignment. First, the use rights of households over land were deemed insecure due to their periodic readjustment in the early years of implementing the Household Contract Responsibility System (HCRS). In the first round of land contracts between the late 1970s and the late 1990s, land was first contracted to households for five years, 10 years or 15 years by different localities. Since collective ownership is actually membership rights which entitle every member equal rights to share collectively owned land (Zhou & Liu, 1988; Kung, 1995, 2000), when new members enter (by birth or marriage) and old members drop out (by death or emigration), land and the revenue from it are subject to periodic reallocation (Rozelle & Li, 1998).

In order to solve the uncertain environment caused by periodical adjustments, the central government regulated that the contract period should be extended to 30 years in the second round of land contracts which started in the late 1990s. This was legalized in the 2002 Land Contract Law. After that, a reform of clarification of individual villagers' rights over rural land was initiated by the central government from top-down. The basic objective was to clarify villagers' rights, to create certain contract relationships between them and village collectives, and to further enhance the efficient circulation of agricultural land. In 2014, the central government announced that the contract period would be further extended 30 years after the expiration of the second-round land contract. Meanwhile, it required the definition of households' contract rights and use rights according to locality-specific contexts. The two main types were boundary clarification of contracted land, and clarifying shares of land held by households without clarifying the physical boundary.

As a result of the institutional reform, the bundles of rights that villagers held have been changing. When HCRS was first adopted, villagers were entitled to: 1) the right to contract land from village collective organizations, the use right over farmland contracted to them under HCRS and the right to residual income from farming; 2) the use right over a small plot of land to build housing for self-use; and 3) the right to benefit from the land when it is requisitioned by the government. Currently, in addition to the abovementioned rights, rural households have the right to transfer farmland to village members and even to outsiders for agricultural uses, and the right to gain rents from the transfer of use rights. Rural housing is now allowed to be leased for rental income.

Second, collective property rights inside villages were deemed as the cadres' rights by many researchers (Brandt et al., 2002; Chen, 2006). In their opinions, the Organization Law of Village Committee released by the central government in 1998 did not clearly define the authority boundary between the bottom-up elected administrative system and the top-down appointed party system in the governance of rural society. There was an intersection between the two systems (Liu et al., 1998; Brandt et al., 2002). According to Chen's (2006) opinion, the village committee operates like a power broker between upper governments and grass-root villagers, leading to a serious principal–agent problem, and whom the village committee will primarily serve is never clearly defined. Alpermann (2001) confirmed that the party branch plays the central role in a village's decision-making in most cases and elected cadres act as the agents for the party. When land is scarce and valuable, cadres have a strong incentive to retain control over the readjustment of land in their own interest (Brandt et al., 2002).

Although village cadres have played a crucial role in rural management and development, the situation has also been transforming due to land rights reassignment in rural China. On the one hand, after the extension of the contract period and the clarification of land rights of the peasants, there are now no periodical readjustments of land contracts which were controlled by the village cadres. A market mechanism is replacing village administrative management in affecting households' decision-making of land uses and transfer of land-use rights. On the other hand, in places where villagers' rights over collective land are defined by allocating shares, the cadre–villager relationship in managing rural land changes to a principal (villagers)–agent (village collective organizations) relationship where the role played by the village cadres has been weakened.

Although the foregoing discussion has provided great insights into the analysis of China's land institutions and pointed out many problems, much attention should be paid to the characteristics of the enforcement of these legal institutions. Zhou contended that it is not the nonfigurative legal land-ownership that imposes factual constraints on the economic behavior of land-related agents. Although formal rules exist, it is also possible that they are not followed by or even known to the agents. If one wants to understand agents' behavior and the resultant outcomes, one has to first determine the "rules-in-use" (North, 1991).

In China's pragmatic reform of "trial-and-error," enforcement by formal institutions may vary in degree across localities and bottom-up creation of new informal institutions is possible. As Liu et al. (1998) noted, with the weakening of state power, many localities have spontaneously initiated various bottom-up institutional innovations based on local factors. Therefore, in-depth exploration of how formal institutions are enforced locally and what new informal institutions have been created are necessary.

5.1.3 The emergence of Land Shareholding Cooperatives (LSC) from the bottom-up throughout China

Decentralization in China's economic reform has promoted the bottom-up initiatives which further lead to institutional changes from the bottom-up according to the locality-specific socio-economic contexts. Situated in the general trend of China's reform towards marketization through a gradual reassignment of property rights (Oi & Walder, 1999), rural reform is also market-oriented based on gradual clarification of rural collective ownership system (Po, 2008). Rural land is an important sphere experiencing great institutional change, but it has always been characterized by ambiguity by law. In the economically dynamic regions which experience rapid industrialization and urbanization, land-related disputes for the distribution of land rents and even the land *per se* tend to occur more frequently. The attempts of rural actors to solve the increasing disputes and promote rural development have led to bottom-up institutional experiments in the redefinition of the property rights over rural land (Fu, 2003). In such a general context, a variety of land-based shareholding cooperative systems have gradually emerged throughout China over time. The cooperative system varies in terms of property rights arrangements and organization structure, according to time-and-locality-specific contexts. The following will elaborate on the emergence and characteristics of land-based cooperatives in PRD and YRD which are two typical representatives.

(1)   The village-based LSCs in PRD

In the early 1990s, the PRD region took the lead in initiating reform of the LSC system. Some villages used experiments under the reform as a way to deal with farmland fragmentation that happened due to the HCRS and the periodical readjustment of land distribution among households caused by population change in the collectives and prosperous industrial development (Fu & Davis, 1998). Under the LSC experiment, land allocated initially by village collectives, except for residential land plots already allocated for villagers' housing, was re-collectivized and managed by the LSC management committee. In return, villagers were given shares of the cooperative, with the number of shares determined by age intervals. Shareholders then shared in the annual dividends allocated by the LSCs.

The focus of LSC was to reassign individuals' rights over collectively owned land without clarifying the land boundaries of each member, which successfully solved the problem of fragmentation of farmland and periodical readjustments. Moreover, LSC reform moved further forward the definition of "collective." LSCs managed the land through leasing some farmland to local villagers initially and later to outsiders, while converting remaining farmland and leasing it for non-agricultural uses. Land rents capture increased collective income which also improved villagers' income.

Cooperatives represent the governance mode of "member ownership, members control, and members benefit" (Ortmann & King, 2007). However, they are also distinguished by how membership is delineated and how benefits are allocated among the members. When the LSC system experiments began, there was not a commonly accepted guideline about how membership should be defined. Dependent upon existing allocation institutions formed under the HCRS, the free-allocation and open-membership governance mode was first adopted. This type of governance was similar to a situation without governance and not good for the sustainable development of collective economy. In search for a better governance structure, two village-led experiments emerged in the early 1990s: one of "closed-membership LSCs" and the other of "open-membership LSCs" where new members must buy access into the LSC.

Initially, closed-membership LSCs were preferred by local government because LSCs were expected to be economic organizations and operate like normal shareholding corporations. Because collective assets were allocated to a fixed group of members and each member's shares were constant, individuals' residual rights were well delineated. Members would have great incentive to contribute to the long-term development of LSCs. Moreover, collective revenue in these closed-membership LSCs was expected to be used chiefly for village welfare and development, while only a very small percentage was allocated as dividends.

Rural land is "owned" by all village members by law. Based on this understanding, all village members should be legitimately entitled to the benefits derived from collectively owned land. Moreover, as LSCs basically relied on land rents for income, the closed-membership governance was increasingly challenged. It gave rise to an equity problem, as new village members moved in by birth and marriage were not entitled to LSC shares, while those members having left villages because of death or migration could still claim collective revenue. Justice among village members became an urgent concern. Open-membership LSCs have been gradually advocated and adopted throughout the PRD region since the early 2000s. In an open-membership LSC, shares, which have been freely allocated to the original members in the previous governance modes of "closed-membership" and "free allocation and open membership," were unchanged. The members, like those married women who did not get shares in the past but still held village *hukou*, could purchase LSC shares for a fee.

The LSC system of this sort was initially considered successful in terms of rural collective income increase, clarification of villagers' rights over collective resources, and sharing land rent appreciation to avoid social conflicts in governments' land acquisition, in spite of some flaws intrinsic to the cooperatives. With an increase in the frequency of land-related conflicts led by extensive urban expansion sponsored by urban governments, many local governments came to see shareholding reforms as a device to mitigate said conflicts (Po, 2008).

(2)   The private LSCs in YRD

In the PRD region, the mode of industrial development at the village scale gradually exhibited apparent disadvantages such as the absence of scale economies and environmental problems. In contrast, urban governments in the YRD region strengthened their control over villages' land conversion and land leasing for non-agricultural uses in the late 1990s, and built deliberately planned development zones to develop industries. This government-dominated urban expansion was usually associated with displacement and marginalization of rural villagers, low compensation, low rural income, and a potentially high rate of land-related disputes.

Kunshan, a county-level city subordinated to Suzhou Municipality, has taken the lead in shareholding reforms in the YRD. Although farmland conversion had been strictly controlled by the city government, the villages were permitted to reclaim and convert unused marsh land into arable land and then develop the increased area of farmland for non-agricultural uses in the late 1990s. However, instead of carrying out development by village collective organizations like that in the PRD, the village collective in Kunshan took the developable land and leased it to cooperatives established by groups of villagers who were interested in profiting from land development (Po, 2008, p. 1615). The members invested money to form a cooperative which then rented the land from village collectives and built factories, workers' dormitories, and commercial spaces for rent. The cooperatives received the differential between the rent paid to village collectives and that gained from leasing of premises, which was shared by the members. New members had to buy access to the cooperative and take part in the land development scheme. In contrast to the village-run LSCs in the PRD, LSCs in the YRD tended to be private.

Because these LSCs generated significant revenues for rural collectives and raised villagers' income, the government began to advocate the LSC system throughout the city in 2003. The government did this by allowing each village to use 3–5% of their farmland for non-agricultural development after cultivation of an equal amount of fallow land to farmland (Po, 2008, p. 1615). In case of land requisition, villages (or villagers) whose land was acquired by government would also be returned a portion of developable land near the requisitioned land plots or in another location near an urban industry cluster. The villages then leased out the returned land to the LSCs for development.

These cooperatives were locally called "*Fuming*" cooperatives, a type of cooperative meant to enrich the people in rural areas and they were advocated throughout the entire YRD region. In Jiangsu Province, by the end of 2009, 1,130 LSCs had been established. While in Suzhou City alone, more than 290,000 households and land of 700,000 Mu were involved in 577 LSCs. By the end of October 2005 in Zhejiang Province of the YRD, there were 502 villages that applied the shareholding co-operative system, in which 14.93 billion Yuan in assets, and 535 thousand members were involved (Tian & Zhu, 2013).

(3) LSCs: welfare-oriented or investment-oriented

A cooperative is "an autonomous association of persons united voluntarily to meet their common economic, social and cultural needs and aspirations through a jointly-owned and democratically-controlled enterprise" (The International Cooperative Alliance, www.ica.coop, cited by Ortmann & King, 2007, p. 41). However, the structures of membership and organization differ very much across cooperatives and lead to a great difference in management styles and the capacity in pursuit of sustainable development. Generally speaking, LSCs in the PRD are village-based and more welfare-oriented, while LSCs in the YRD tend to be owned and managed by "private" groups of villagers and are more investment-oriented.

The membership structures are different between the two kinds of LSCs. In PRD, members with agricultural *hukou* in the village are naturally the shareholders of the LSCs. All collective resources are managed by the cooperative committee and collective assets belong to all members of the LSCs. The LSC reform in the PRD does not alter the collective basis of the economy. However, in the YRD, the LSCs are established by individual villagers on the principle of voluntary participation, and only the rent gap belongs to LSC members.

There is also a great difference in the definition of individual's rights in the LSCs of the two regions. In the PRD, shares allocated to shareholders by age instead of contribution are only an instrument to distribute annual dividends. No more than 60% of annual revenue is allocated according to shares, while the rest is retained by the LSC for members' welfare based on the egalitarian principle and is reinvested in public facilities which benefit all members. It is apparent that how much collective assets a shareholder owns is not clearly defined. That is to say, shareholders' residual claimant rights are ambiguous. In contrast, in the YRD, LSC members' rights are clearly defined. Land management and transfer of collectively owned land are not solely dominated by the villages, but are defined by contracts between a village and the LSCs (Po, 2008). The LSCs' duty is specialized, only engaging in the management of premises which belong to the shareholders and every member invests to join the cooperative. The number of shares a shareholder holds is determined by how much he/she invests in the cooperative.

To summarize, as Po (2008, pp. 1614–1619) pointed out, LSC reform in the PRD places the village as a collective on center stage, further consolidating the collective in rural development. Different from LSC in the PRD, LSC reform in the YRD is not intended to bind villagers more tightly with collective economies. It aims to promote local entrepreneurial efforts through the volunteer participation of villagers in market-driven land and property development, not a welfare system.

5.2 Land markets and stakeholders during peri-urbanization

5.2.1 Land markets structured by the changing land institutions

(1)   Formal land markets under the legal institutions

Markets are structured by institutions. Shaped by the framework of dual landownership, China's land market is composed of three sectors: urban land market, rural land market, and the market for the conversion of land from rural to urban. These markets emerge and develop with the transformation of the perception of land from being a means of production to being an asset with investment value.

Urban land markets emerged after the permission of separating land-use rights from land ownership was granted in 1988 through the amendment of the 1982 Constitution. In the 1982 Constitution, Article 10 that states "no organizations or individual may appropriate, buy, sell, or unlawfully transfer land in other ways" was amended to "the right to use land may be assigned to land users in accordance with the provisions of law."Before 1988, state-owned urban land was considered as a means of production, which could only be administratively allocated to users by the state. After 1988, the commercialization of land-use rights had been legitimized to complement an emerging market economy. Urban land can now be leased to developers or users for a fixed period of time after a payment of rent in a lump sum to the state (Zhu, 1994, 1999, 2005; Ng & Xu, 2000). Moreover, land-use rights can be freely transferred among land users in a secondary market. According to Lin and Ho (2005), the amendment introduced a new market track into China's land system and, together with the traditional plan track, gave rise to a dual-track land management system. Gradually, the market track replaced the plan track as the dominant means for the transaction of urban land.

Different from state-owned land, rural collectively owned land has always been treated as a means of production. However, some scholars have observed that "markets" also exist in the rural area. According to Ho and Lin (2003), the market can be divided into the agricultural land sector and the construction land sector. Since farmland was contracted to individual households under the HCRS, it was permitted to circulate among households inside villages by the 1988 LML and the scope of circulation was extended to outsiders in the 1999 LML. As to the non-agricultural land sector, circulation has always been restricted by law. The conversion of agricultural land into non-agricultural use was permitted only when it was for self-use by the households to build homes or by the rural collectives to launch TVEs and build public facilities (Article 43, LML, 1999). According to Article 60 of the 1999 LML, rural collectives as the owner of land can launch a collective enterprise independently or start joint-ventures with other "state agents" and individuals. Also, rural collectives can use the ownership certificates of buildings on legally converted land and the certificates of the use rights of construction land as collateral. However, Ho and Lin (2003) mention that few banks are willing to accept such collateral, because it cannot be circulated in the commercial secondary market, rendering the certificates with little value.

For the conversion of rural land to urban land, urban governments have the authority to acquire rural land after paying the rural collective an expropriation fee in a lump sum that consists of compensation for the land loss, a resettlement allowance for displaced peasants, and compensation for lost crops. The fee is set at several times the average annual agricultural output for the previous three years. The maximum lump sum payout is regulated at less than 30 times the previous three-year average. According to some scholars, this low compensation amount shows the intrinsic attribute of rural land as a means of production instead of as an economic asset to the nominal owner of rural collectives (Wen & Zhu, 1996; Zhu & Hu, 2009).

(2)   Informal markets and its formalization in the rural land sectors

Besides the formal markets, some illegal markets, or "black markets," exist in urban or rural areas. In addition to the black markets in the urban land sector, there were at least two "black markets" in the rural land sector. The first type of "black market" involves informal house-leasing markets in urbanizing villages. According to the LML (2004), village housing lots are meant for personal use and not considered as assets for lease or for sale. Nevertheless, with the influx of a large number of migrant workers into the city, renting space from village houses becomes a convenient and economic option. In order to accommodate more migrants, the houses are usually built larger than the legally allowed plot size.

The second type of "black market" is land conversion for non-agricultural uses without proper approval from the urban government. Many scholars have attributed the existence of such a "black market" to the ambiguous and incomplete definition of land development rights between the "collective" and the government, and the gap between compensation and conveyance fees (Zhu & Hu, 2009). For instance, before the implementation of the LSC system, it was possible for urban governments to seize lands from villages. The situation has drastically changed under the LSC system, as LSCs have managed to strengthen their holding of collective land by stipulating that an agreement from more than two-thirds of cooperative members is required in the case of alternation of cooperative assets where land stock is the major component, and this has significantly curbed the power of the urban state in land acquisition. After obtaining agreement from cooperative members, the LSCs convert farmland and lease the land to enterprises for land rents after obtaining approval of local government under the guise of establishing joint-ventures, or without any approval at all.

5.2.2 Stakeholders and their interaction during peri-urban development

In the land development process, there are many actors and interests. Generally speaking, on the supply side, the governments at different levels as agents of the central state, and village collectives as the nominal owner of rural land, are the key players. In the demand side, tenants and migrants are the two crucial agents who interact with the supply-side actors to affect transformation in peri-urban areas.

(1)   Governments at different hierarchies

With the reform and opening, the central state has become development-oriented in the process of marketization. Decentralization has become the primary instrument of the central state to fulfill its objective of turning local governments into "local developmental states." Devolution occurs in the central–local intergovernmental power structure to give localities more freedom in making investment decisions and managing local growth (Zhu, 2005). The shift of power downward is well reflected by the central state's sharing with local governments the fiscal revenue which was strictly controlled in the planned system by the former (Huang, 1996). Being residual claimants' of fiscal revenue, local governments have a strong incentive to pursue local economic development (Qian, 2000). Therefore, advancing development strategies that can stimulate local growth and expand fiscal capacity has become an indispensable goal of local governments (Wong, 1992; Qian & Weingast, 1997).

However, within a principal–agents' relationship, the performance and choices of local governments depend on what kinds of incentive structure the central state creates for them. Such central–local interrelationship is clearly shown in the balance between decentralization and concentration in terms of fiscal system reform. As early as 1984, a fiscal-contracting system was used to motivate local governments to develop their local economies. In the land system, fiscal revenue was divided into central fixed revenue (e.g., revenue from centrally owned state enterprises), local fixed revenue (e.g., revenue from locally owned enterprises), and a central–local shared part (e.g., enterprises under dual ownership) (Lin & Liu, 2000). Under this arrangement, local governments were greatly motivated to develop local enterprises such as TVEs and even maneuver to shift some revenue out of the levy shared by the central state (Che & Qian, 1998; World Bank, 2002).

Because of the deep involvement of local governments into the operation of enterprises, local governments were known as local state corporations. Although fiscal decentralization provides great incentive to local agents, competition between the central government and localities for revenues compromised the central government's capacity in central coordination and regional redistribution, which the central state did not wish to see (Oi, 1992, 1995; Wong et al., 1995; Zhu, 2005). On the one hand, the state started to grasp its political central control through the reshuffling of provincial officials. On the other hand, a new round of fiscal reform was initiated in the mid-1990s.

In the new tax-sharing system, the central state and localities share taxes by categories. According to Tao et al. (2009), such a system clarifies the central–local fiscal boundary and also facilitates the center to concentrate power. With the tightening of localities' fiscal power by the central state, land becomes an important source of the local governments' fiscal revenue. Land acquisition and conveyance not only creates significant extra-budgetary income, but also boosts economic growth and broadens the tax-base in the long run (Po, 2008). Relying greatly on land revenues, local governments have become aggressive in rural land conversion for urban uses (Lin & Ho, 2005).

Among the hierarchical local governments, the town government as a power-broker between its upper governments and village collectives has been considered as having a great impact on land conversion. Transforming from an organization governing rural society in the planned system to the bottom of the state bureaucracy in the post-reform era, its power is highly uncertain in dealing with its upper state system and the autonomous villages below them (Hsing, 2006). In this case, "town officials would strategize to maximize their control over village land and profit from the booming land leasing market in the quickly industrializing and urbanizing areas" (Hsing, 2006, p. 103).

Being development-oriented and land-dependent, urban governments in the hierarchy have formed various kinds of coalitions to capture rural land. In Guo's (2001, p. 426) case study, it is the alliance between the urban (county and town) governments and village cadres that leads to land expropriation and conflicts. Through this coalition, the urban government captured 60–70% of total land-related income, whereas the village collective received 25–30% and villagers 5–10%. In the game of land development, the real winners appear to be municipal, county, and town governments, whereas the losers are, ironically, the central state and individual peasants, original owners of China's land. However, does the above literature describe a general situation or just some special cases in which rural collectives are disadvantaged and passive in land conversion? It has been found that in rapidly urbanizing regions, many rural collectives have managed to capture land rents, which will be shown in the next section.

(2) Village collective organizations

Rural villages in China are an autonomous unit in governing rural society and advancing the village economy. Autonomy means self-sufficiency. Historically, due to farmland resources made increasingly scarce by the continuous population growth, villages in China became more territorial in the face of constricted economic opportunities and perennial disorder. Self-sufficiency in the setting of extreme land scarcity caused clan-based villages to become introverted. In terms of economic development, every village had to rely on their own resources and labor forces, with land being the most important part. In modern China, village autonomy has been legitimized by law. Without government intervention, village development would be self-contained.

A growing number of empirical studies have suggested that the black (or informal) land market is prevalent in the rapidly industrialized coastal regions (Lin & Ho, 2005; Po, 2008, 2011; Zhu & Hu, 2009). For example, it was estimated that 50% of total construction land in the PRD is attributed to land conversion by rural collectives without landownership change (Guangdong-Hongkong Information Daily, 17 November, 2003). In these areas, village collectives have developed a variety of strategies to lease their collective land to outsiders for non-agricultural uses. They have spontaneously violated the formal institutions, undermining the notion that land for urban uses must be acquired through the state acquisition process (Po, 2008).

It has also been observed that active participation of village collectives in land conversion is embedded into a series of bottom-up institutional innovations. One of the most significant innovations is the land-based rural shareholding reform which while initiated in Nanhai, Guangdong, later spread in various forms to many locales in China (Fu & Davis, 1998; Po, 2008). In this reform, village collectives restructured their organizations and redefined villagers' rights over collective assets in which land is the most important part (Po, 2008). Traditional village governance has been transformed into a structure centering around LSCs. Rural land has once again been collectivized under the operation of cooperatives. Village collectives play different roles in land conversion in different social-economic contexts, and they interact with the government in different hierarchies, shaping different models of peri-urbanization.

(3)   Land tenants and inflow migrants

With collectively owned land, villagers are entitled to land lots for building their family houses. Village housing lots are meant for personal use and are not considered as assets for lease or sale. Housing lots can, however, be transferred and inherited by immediate family members. Some village households may inherit lots from their parents and thus possess more than one housing lot (Ministry of Construction, People's Republic of China, 2008).

In comparison with urban households, rural households have an affordable housing option because housing land in rural villages is given free of charge. However, in poor rural villages, building a proper house is one of the greatest financial burdens on peasant families. According to Gao (1999), a simple single-storey house with a wooden frame structure and mud clay walls would cost ten years' income of a male adult peasant. Before the rural reform began in the early 1980s, rural housing was of poor quality. It was the TVEs that began the transformation of rural life. Improved village houses were probably the first sign of rural change. In the PRD region since the 1980s, revenues from the growing TVEs were spent on village infrastructure, and wage incomes of village workers were used to rebuild and refurbish their houses.

Low-income urban residents with *hukou* may have access to affordable housing provided by the municipal government if they cannot afford commodity housing in the market. Meanwhile, migrant workers are not entitled to any government-sponsored affordable housing schemes in the host cities, and they can only find accommodation from the market. Because of their low wages and heavy responsibilities of supporting children's education and maintaining elderly parents, migrant workers do not have a budget for decent housing. Renting space from village houses becomes a convenient and economic option as they are working in the factories located in the villages.

Rural reforms and industrialization have greatly enhanced the economic well-being of villagers in the dynamically industrializing townships. Having perceived the potentially substantial demand for housing from migrant workers, many villagers rebuild their houses to a size much greater than their own actual need in order to provide additional space for leasing.

Given the nature of rural housing as collective welfare, village housing is meant only for personal use and its leasing is not officially permitted. However, because of the contribution of migrant workers to the local economy and the incapability of the local government to provide them with affordable housing, informal village housing leasing is tacitly allowed. A large informal leasing market of village housing has appeared, and so have villagers as the housing landlord/or renter class.

5.3 Peri-urbanization led by autonomous LSCs under institutional uncertainty: a case of Nanhaiin PRD

Nanhai is located at the fringe of both Guangzhou and Foshan, to the immediate west of the former, and to the north of the city proper of the latter (Figure 5.1). Nanhai used to be a rural county, but became an urban district annexed to the Foshan municipality in 2003. With a very good location, Nanhai received substantial outside investments and became a dynamically growing district in the PRD. From 1982 to 2010, its gross economic output increased from 800 million Yuan to 179.67 billion Yuan, driven mainly by manufacturing which contributed at least 50% to the total economic output.

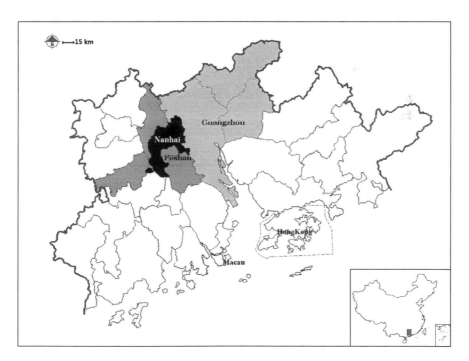

Figure 5.1 Location of Nanhai in the Pearl River Delta.
Source: drawn by the authors

The total Nanhai population in 2010 was 2.59 million, nearly 3.14 times of that in 1982, among which 45.9% (1.19 million) were residents with *hukou* and the remaining 54.1% (1.40 million) were migrant workers without *hukou*. Although 93.3%of rural laborers had left the field by 2010, those with local rural *hukou* did not decrease. Instead, their population increased from 666,900 in 1982 to 757,700 in 2010 (NHAB, 1982; NHBDPS, 1998–2011). In the meantime, the proportion of local urban residents of the total population only increased from 19.3% in 1982 to 36.3% in 2010; 77.3% of immigrants were registered in villages in 2010, with a population of 1,083,200, and 71.0% of the Nanhai population living in villages. Although most residents have engaged in non-agricultural jobs, the ways of village life remain.

With rapid economic development, Nanhai experienced extensive urban expansion between the early 1980s and the late 2010s. Its construction land area increased from114.5 km^2 in 1990 to 568.8 km^2 in 2008 (Figure 5.2). The construction land as a percentage in the total territory was 53% in 2008. Further analyses have revealed that the rapid development in Nanhai seems to have been driven by diverse landed interests and two modes of land development (Figure 5.3). One of these modes was land development initiated by urban state, with landownership changing from the collective to the state. The other was the bottom-up rural industrialization initiated by the village collectives, with collective landownership unchanged. In 2008, the former contributed 34.8% to the total construction land, and the latter accounted for 65.2%.

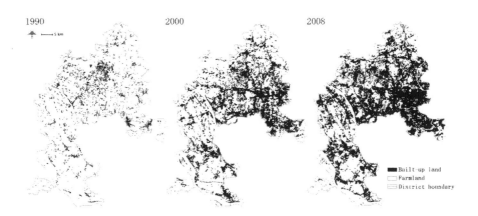

Figure 5.2 Growth of the built-up area in Nanhai, 1990, 2000, and 2008.

Source: derived from satellite remote sensing maps at: datamirror.csdb.cn/admin/introLandsat.jsp

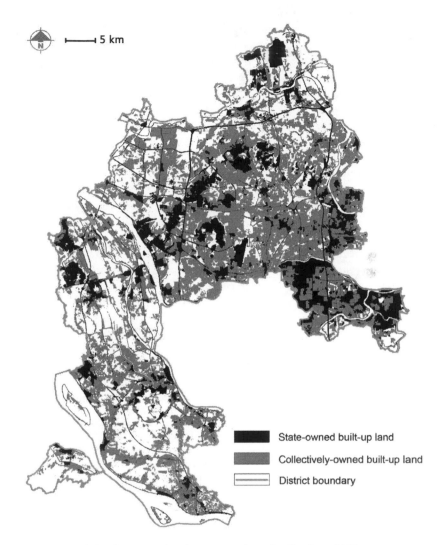

Figure 5.3 Land developments by the state and rural collectives, 2008.
Source: Archives of Nanhai Bureau of Land and Resources

5.3.1 Bottom-up change of land institutions during rapid rural industrialization

(1)   HCRS and rural industrialization driven by TVEs

Although the HCRS greatly increased agricultural productivity in the early stage, it gradually exposed two weaknesses: the land fragmentation and inappropriate land allocation in the absence of rural land markets. First, under the HCRS, farmland was contracted to individual households for a period of time (5–10 years initially) in the early 1980s. Such land distribution was implemented based on an egalitarian principle which required equality in terms of both land quantity and quality (Kung, 1995, 2000). Therefore, more often than not, a household would have several spatially dispersed plots of farmland which would vary in the degree of fertility (Wang, 1994). This resulted in the fragmentation of farmland, which was more serious in Nanhai because of its severe scarcity of land. Farmland per capita in Nanhai was only 0.6 mu in the early 1990s, one-third of the national average (Wang, 1994). Moreover, land distribution had to be readjusted periodically because of the change in both village population and the quantity of farmland. Each readjustment exacerbated the problem of land fragmentation which does not allow for economies of scale. As a result, further improvement in agricultural production was seriously hampered due to the extreme piecemeal land subdivisions (Fu & Davis, 1998).

Second, in Nanhai and other rapidly industrializing cities in the PRD, non-farm jobs have been abundant since the 1980s. Instead of farming barren land, an increasing number of villagers went to factories for better salary. By 1990, 20% of rural laborers in Nanhai had already been engaged in non-agricultural employment (Hou & Zeng, 1991). Those with non-farm jobs had to retain part-time involvement in cultivation and at times let the land lie idle. On the one hand, they had to submit a regulated quota of agricultural products to the government. On the other hand, land transfer between households was forbidden and land was not allowed to be sub-leased to outside villagers (Fu & Davis, 1998; Yao, 2000). Therefore, there was a dilemma in which the demand for land by the farming specialists and idle farmland co-existed, which requires institutional change.

In the early 1980s, following Guangdong's economic reforms which encouraged the participation of various economic actors, the Nanhai government initiated a ground-breaking policy guideline to develop a decentralized economy. This was done by engaging six engines: county-owned enterprises (state-owned); enterprises owned by townships, administrative villages and village groups (collectively owned); privately owned enterprises and joint ventures with foreign investors (www.wzg.net.cn/city/toRead.php?vid=324; accessed on June 30, 2012). This created the well-known "Nanhai model," characterized by mobilizing all sorts of market actors to create a decentralized economy. Encouraged and supported by this policy, collectively owned enterprises grew quickly in Nanhai, with its contribution to the total industrial output rising from 17% in 1978 to 80.5% in 1995. Village enterprises, in particular, contributed 47.2% to the total industrial output in 1995, rising from 2% in 1984 (NHBDPS, 1998–2011).

(2)   Natural-village-based LSCs

The model of Nanhai's decentralized economic development stimulated the bottom-up institutional innovation. In order to conquer the aforementioned dilemma of inappropriate allocation of land resources, the LSC system was first created by some villages in the early 1990s. Under this system, all land parcels except family housing plots in the village were re-collectivized and operated by the LSC management committee. In return, village members could hold the shares of LSCs, and shares were distributed to cooperative members equally based on a formula which took into account members' past contribution to the collective assets and could not be transferred or sold. More than 40% of the collective income from farmland leasing, village enterprises and properties, would be retained by the LSC for future development, public and welfare expenses. The remaining part was distributed to the shareholders. The LSC management committee was elected by its members based on the principle of "one member, one vote," which was different from the normal business undertakings that were investor-oriented and professionally managed.

As the basic social unit in rural China, natural villages have been a very stable and steady institution throughout Chinese history (Gao, 1999). Based on clan and lineage, there was a close solidarity and coordination within the rural community managed by the village gentries before the 1949 revolution. The founding of the People' Republic in 1949 did not change the nature of China's natural villages as a basic socio-economic unit. In the following People's Communes, although decision-making in rural production was made by the Communes, the teams (natural villages) were the basic units of rural production and land uses. According to the survey of Ministry of Agriculture of China (1993), under the HCRS, land distribution to village households in rural China was mainly implemented by the team (natural villages), with the minority occurring at the administrative villages. Though the Organic Law of Village Committees has officially promulgated that administrative villages should be the basic autonomous socio-economic units in rural regions, natural villages in Nanhai are actually the basic units, usually consisting of members of the same clan. Most LSCs in Nanhai are established at the natural-village level.

By 2010, 2,063 LSCs had been established in Nanhai, in which only 13 LSCs had been set up at the level of administrative villages, consisting of several natural villages. The remaining 211 administrative villages established 2,050 LSCs at both the administrative-village (211) and natural-village levels (1,839). All members of the administrative village were the principals of the administrative-village-based LSCs which allocated dividends or/and welfare to them. The natural-village-based LSCs managed the land which specifically belonged to the natural village members and was responsible for their allocation of benefits. Regardless of the levels of villages to establish LSCs, LSCs were reckoned as both social and economic autonomous organizations, as villagers are the natural members of LSCs. The role of the LSC committee was to develop the rural economy and increase shareholders' income and welfare.

(3)   Land leasing for non-agricultural uses as informal institution

After the establishment of the LSC system, the LSCs initially ran the village enterprises and leased farmland to village households for revenue. However, since the late 2000s, the collective enterprises (TVEs) have faced increasingly fierce competition from inward foreign and private enterprises which had industrial expertise and understood modern business management (Che & Qian, 1998; Pei, 2002). As a result, the TVEs gradually withered, while privately owned enterprises and joint-ventures flourished. The TVEs' share of total industrial output declined drastically from 50.2% in 1998 to 2.9% in 2008, while the ratio of foreign and private sectors jumped from 40.9% to 96.4% (NHBDPS, 1998–2011).

The decay of collectively owned enterprises finally led to the reform of their ownership in the mid-1990s, mainly in terms of restructuring into private enterprises (Park & Shen, 2003). To facilitate ownership transformation, special and temporary policies were issued to adjust the property rights over construction lands occupied by the enterprises. The construction land could be acquired by the state or be directly leased to the transformed enterprises by the collective without a landownership change (Ho & Lin, 2003). However, collectives' construction land leasing was only applicable to special cases involving transformation of TVEs at that time. With the dropping out of collectively owned enterprises, LSCs' economies were in peril of a depression, which promoted informal institutional change from below. Aspiring to lead the way of industrializing villages, LSCs took the role of landlords, converting agricultural land and then leasing it to private enterprises.

LSCs' conversion of agricultural land for non-agricultural leasing was informal, or strictly speaking, illegal. As mentioned in Section 5.1, the official doctrines of collective landownership regulate that only the urban government has the right to develop land to accommodate non-agricultural tenants and rural land can only be developed for self uses after the approval of urban governments, such as TVEs and villagers' housing. Accordingly, by law, LSCs' land conversion for leasing occurred under three situations:

1.   LSCs may have received the approval of the urban states in application for land conversion under the guise of initiating joint-ventures,[1] while it is too costly for the government to verify such an application.
2.   LSCs may have never applied for approval from the urban state. It has been revealed that 7,408 ha of land was converted in this form between 1997 and 2004, accounting for 34.9% of the total construction land growth during the same period (NHBDPS, 1998–2011).
3.   When implementing the LSC system, it has been regulated that an agreement from two-thirds of the cooperative members is required in the case of alteration of cooperative resources where land stock is the major component. This strengthens LSCs' hold on collective land and curbs the power of the urban state in land acquisition. Moreover, the local government

must consider the sustainable economic development of the LSCs to consolidate their reform achievement. LSCs' land leasing may be tacitly agreed to by the local urban government. However, from a legal perspective, LSCs' land conversion for non-agricultural uses is informal.

5.3.2 Land rents competition under institutional uncertainty: rapid and fragmented non-agricultural development led by LSCs

(1) Land rents competition between LSCs and urban state under uncertainty

Land leasing for rentals is significant for collective economy and villagers' welfare. In 2006, 2009, and 2011, rental income from non-agricultural land leasing in Nanhai accounted for 84.7%, 69.1%, and 74.5% of the LSCs' total revenues, respectively NHDRA (2006–2012). According to the requirements of the Nanhai urban government for distribution of collective income of LSCs, no more than 60% of total collective revenue can be distributed to shareholders, and the remaining revenue must be retained for future village development. In reality, on average, 41.6% of the LSC income was distributed as dividends, and 19.9% was used for providing social services and welfare such as health, education, and elderly pension to shareholders between 2006 and 2011 (NHDRA, 2006–2012). Until 2012, Nanhai had converted 61.7% of its territory to industrial land uses, and the dividends from collective economy and household housing rentals represented 35.3% and 15.7% of the total household income, respectively.

As agents of the shareholders, LSCs tend to make improvement of villagers' dividend allocation and welfare its first priority. Although land development rights are defined to be owned by urban state by law, the LSCs have managed to gain more rights over controlled landholdings than their predecessors (village collectives). In order to increase collective revenue, the LSCs actively engage in farmland conversion and leasing for non-agricultural uses, scrambling for as much land as possible and to make it available for leasing. Under this circumstance, LSCs have transformed from organizations of economic production to entities of landlord economies, relying on rental incomes from industrial land leasing (Zhu & Guo, 2014, 2015).

Apparently, the informal land conversion of LSCs is motivated by the capture of land rent differentials. Despite the tacit agreement of local urban governments, land leasing for non-agricultural uses is an informal activity as it is not formally recognized by the central state. As a result, land development rights are ambiguously defined between the LSCs and local urban states. Land rents are left in the public domain, and up for grabs, as mentioned by Barzel (1989). Wen (1998) found that compensation paid by government to rural landholders for the appropriation of one mu of farmland was only 28,000 Yuan in the early of 1990s. Our investigation in the villages of Nanhai revealed that rental income from one mu of industrial land could be as high as 70,000 Yuan. Impending land acquisition by the state makes villages anxious and restless. Therefore, the LSCs have every incentive to convert farmland to non-agricultural uses under their control.

Given the fierce competition for land development, urban government has to opportunistically negotiate with every LSC for land acquisition for government-sponsored projects. The Nanhai district government has been an active actor in the development of development zones such as industrial park. Nevertheless, in the informal institution of land leasing, the rights of the state to develop agricultural land for non-agricultural uses have been encroached upon by the LSCs. It has been difficult for urban state to acquire land since the extensive establishment of LSCs in the mid-1990s. This is evidenced by the development of Nanhai's urban and town centers. Land use statistics show that construction land of urban and town centers increased from 177.7ha in 1988 to 4,129.9 ha in 1996, while its ratio of total construction land increased from 1.4% to 12.9%. However, in 2008, the construction land of urban and town centers was only 4,999.5 ha, and its ratio of total construction land area decreased to 8.6%.

Additionally, conversion of an entire village into an urban area by the urban government is becoming increasingly difficult, as villagers would be deprived of their land. Opportunistic land acquisition from villages based on negotiation led to the dispersion of urban projects (state-owned land) spatially with 789 patches across 181 administrative villages (Figure 5.4).

(2) Self-contained development of autonomous LSCs: rapid and fragmented land development

The LSCs, with an area of about 41ha on average, carried out self-contained land development within a small territory, leading to rapid and fragmented land development. First, land leasing for non-agricultural uses is informal. Being cultural-specific, informal institutions may not be sufficient to provide order in the absence of formal institutions when society becomes diverse and heterogeneous (World Bank, 2002). There is no existence of long-term certainty for LSCs' informal land conversion and security of informally acquired land rights. When uncertainty occurs, it prevents one from making rational and long-term planning decisions. Instead, uncertainty can induce disordered competition for short-term benefits, especially within the setting of a heterogeneous society. As a result, LSCs tend to concern themselves more with short-term profits.

In order to strategically preempt the urban state's land acquisition at a relatively low compensation, LSCs hastily covert the farmland in under their control as fast as they can. This contributed, to a great extent, to the rapid land conversion after the establishment of LSCs in Nanhai. It is evidenced that 58,418.2 ha of land had been converted to construction land by 2008, accounting for 54.4% of the total land area of Nanhai. State-owned construction land only accounted for one-third of the total construction land, and the remaining two-thirds belonged to rural construction land

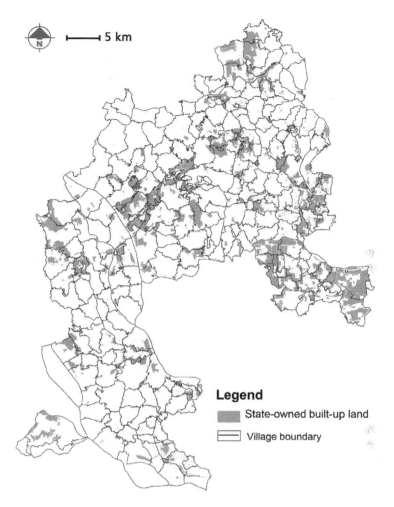

Figure 5.4 Dispersed urban state-led land developments across villages, 2008.
Source: Archives of Nanhai Bureau of Land and Resources.

which was overwhelmingly used by rural housing and industries. Industrial land amounted to 57% of the total construction land and rural industries took three-quarters of it, showing industrialization was mainly driven by rural sectors in Nanhai. The fast land conversion by LSCs was seen in agricultural land being converted to industrial land at a rate of 1,280.3 ha per year during 1988–2000, with this rate increasing to 1,941.5 ha per year during 2000–2008 (NHLRB, 2008).

Second, LSCs' self-contained land development at a small scale, together with government's opportunistic land acquisition, led to landscape fragmentation. As

Table 5.1 and Figure 5.2 show, with Nanhai district as a whole landscape, the number of farmland patches increased from 77 in 1990 to 885 in 2008, and the mean patch size gradually fell from 1,245ha to 63ha per patch, indicating that farmland was gradually cut into pieces by construction land. Also, with constriction projects penetrating into rural areas, farmland and construction land was extensively piecemealed and spatially intermingled. The contagion index which measures the extent to which land uses are aggregated or fragmented, with landscape aggregated at high values and otherwise dispersed (O'Neill et al., 1988), gradually dropped over time. This shows that the degree of intermingling of farmland and construction increased and Nanhai's landscape became increasingly fragmented.

Further analysis at the village scale reveals how the landscape was fragmented and different land uses were mixed at the micro landscape level. In 2008, there were 1,868 patches of collective industrial land, 1,862 patches of village residential land, in addition to the aforementioned 789 patches of state-owned land. The average patch area was 18 ha for collectively owned industrial land and 25.39 ha for state-owned land. If measured at the village scale, the Mean Patch Size (MPS) was reduced, as a spatial connected patch would be divided into (or) among different villages. It is shown that collective industrial land was dispersed across 221 administrative villages (the total number was 224) (Table 5.2 and Figure 5.5). On average, an administrative village had eight spatially dispersed industrial land patches. Across all 221 administrative villages, the mean of the MPS of collective industrial land at the administrative village level was 15.79ha. As for administrative villages, 223 of them had residential land. The average number of village residential land patches was six, and the mean value was 12.24ha. Because of the opportunistic land expropriation of urban governments, state-owned land was also dispersed across 181 administrative villages. The average patch number was four and the mean value of patch size across all villages was 16.83 ha. Farmland existed in 215 administrative villages. On average, an administrative village had nine patches of farmland, and the MPS was 34.45ha.

Table 5.1 Patch analysis of farmland and contagion index, Nanhai

Year	Number of Patches (Units, ≥1)	Mean Patch Size (Hectare, >0)	Contagion Index (0–100)
1990	77	1,245.46	62.70
1995	314	260.68	40.78
2000	448	164.82	38.02
2005	680	96.76	33.91
2008	885	63.33	30.28

Source: the estimation are based on data from satellite remote sensing maps at datamirror. csdb.cn/admin/introLandsat.jsp.

Table 5.2 Indices of four types of land uses at the administrative village level, 2008

Land-use Types	Land-use Distribution Across Administrative Villages (Village Numbers)	Mean of Administrative Village-level Patch Numbers	Mean of Administrative Village-level Average Patch Size (ha)
Collectively owned industrial land	221	8.45	15.79
Village residential land	223	5.56	12.24
State-owned land	181	4.36	16.83
Farmland	215	8.66	34.45
Total	223	–	–

Source: Nanhai Bureau of Land and Resource

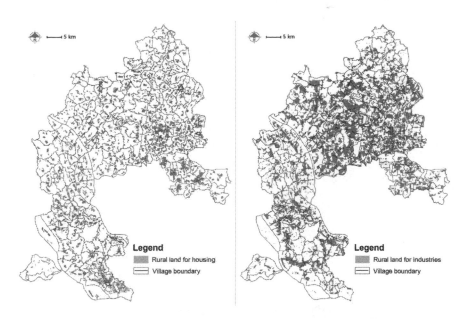

Figure 5.5 Fragmented spatial distribution of residential and industrial land areas in Nanhai, 2008

Source: Nanhai Bureau of Land and Resource

5.3.3   Underproductive utilization of land in LSCs and its ossification

(1)   Problems intrinsic to the organization of cooperatives

A cooperative, as an autonomous association of members, may be able to serve its members' needs in certain social contexts. However, as an organization, it has intrinsic problems such as free-riders, horizon problem, and organizational control which are due to flaws in the assignment of property rights (Cook, 1995; Royer, 1995). Free-rider problems occur when property rights are not transferable and not clearly delineated in an open membership cooperative. Those who contribute less to collective assets (e.g., new members) can obtain nearly the same rights with or even more than those who contribute more (e.g., existing members), causing a dilution of returns to the latter.

The horizon problem arises when the productive life of collective resources is longer than the time span a shareholder can claim on the dividends generated, as the shareholder may leave or die (Vitaliano, 1983; Porter & Scully, 1987). The horizon problem is derived from the problem of restrictions on transferability of residual claimant rights. Therefore, there is a disincentive for members to contribute to the long-term investment and an incentive for members to maximize short-term dividends.

The lack of organizational control is a great difficulty in carrying out efficient management. This comes about because, on the one hand, management of the cooperative is built on a democratic member-controlled system (one member one vote), and the transaction costs in reaching management decisions can become very high. Moreover, the costs of group decision-making increase with the size and diversity of the cooperative (Staatz, 1987). On the other hand, as the contribution is unequal to the return, the manager could be disincentivized to invest time and effort to more efficiently manage the cooperative. In the same vein, investments to supervise managers' decision-making *per se* become public goods to the members, which very easily generates a free-rider problem. Therefore, members could be disincentivized to have efficient supervision. The control problem may finally lead to the incapability of cooperative management to timely react to market opportunities, leading to management inaction.

(2)   LSCs as a welfare organization: the cooperative problems and short-termism

Since the late 1990s, the open membership model has been advocated by the Nanhai district in establishing LSCs (NHDRA, 2006–2012). When the LSCs were initially established, the original members were naturally the members of the LSCs which allocated them free shares. Since that time, new members, through marriage or birth, as long as their *hukou* was in the village, also had rights to buy access into the LSC. Buying, instead of being given shares free of charge, was based on the perception that they had not contributed to the village assets which were created using the retained collective revenue that belonged to original members.

First, there is a pre-determined formula for the purchases of shares, which is age-based following China's old tradition that society needs to take care of aged members. Age intervals are pre-determined first, and each person can only buy the number of shares in the interval to which he/she belongs. When the members advance to a higher age interval, he/she can buy additional shares. Also, if members drop village *hukou* because of death or moving away, they can sell their shares. Nevertheless, the transactions of buying and selling can only occur between members and the LSCs, and the shares are not allowed to transact freely among members, let alone outsiders.

The LSCs with the abovementioned characteristics also suffer from intrinsic problems due to their unclear definition of residual claimant rights of members and problematic governance structure. First, the free-rider problem exists for two reasons: 1) the number of shares a member can buy is fixed according to his/her age. Under the LSCs, village members no longer participate in the daily operation of LSCs. Therefore, age may not represent a members' real contribution to the collective resource. This leads to a situation where senior members who have contributed less to the collective revenue receive more shares than those who really contribute to the collective income, like managers. Our field investigation in an LSC, called PN, shows that a shareholder as a 60-yearold can hold up to 3 shares, while a 20-year-old member can hold only one share. Locals consider this scheme as welfare for the village elderly. In essence, the elderly free-rides over the real contributors, possibly disincentivizing the latter from making contributions.

2) As shares are not allowed to be freely transacted, the actual transaction price cannot reflect the real value which should take into account future potential income flow. The share prices are often determined by the LSCs arbitrarily. In PN LSC in 2009, shares could be purchased and sold at 16,000 Yuan per share while the dividend was 1,820 Yuan per share. At that rate, the "investment" could be recouped in nine years. In another LSC called SB, the share price in 2011 was only twice the yearly dividend. New members and those buying additional shares purchase shares at a low price, resulting in the dilution of profits of existing members who were still not qualified to buy. The former free-ride on the latter, and the latter thus prefer short-term dividends distribution.

Second, the horizon problem is expected in the LSCs, as the revenue of land rents tends to be endless while a lifespan is limited, especially when the share price is too low for those who leave the LSC by death or moving away to receive full compensation. In sum, land shares are more about short-term welfare than investment, as claimed by Po (2008).

Third, LSCs also suffer from the organizational control problem which makes their daily management conservative. As LSC management committee members are elected by shareholders through "one member one vote" every three years, a management committee administrates the village economy with major decisions made in general meetings attended by at least two-thirds of total shareholders and approved by the two-thirds majority of those present (Po, 2008). This election and decision-making system determines an "agent–principal" relationship between the committee members and shareholders. It seems as though LSCs are actually controlled by the shareholders. However, in reality, as there is ambiguity of residual claimant rights, the supervision of the shareholders *per se* is public goods, which faces the problem of free-riders. As a result, the shareholders often have a disincentive to make great efforts to supervise the managers. Meanwhile, managers as the agents of shareholders contribute more to the collective assets, but cannot receive a corresponding return. There is also a great impetus for managers to prioritize their position instead of taking risks to capture market opportunities. The authors' interviews with 32 LSC committee leaders show that the most important aspect of LSCs to exist is providing members with annual dividends and if there is no increase in those dividends, there definitely must be no financial losses for those members.

Shares of LSCs tend to be welfare instead of an asset with market value, leading to the short-termism of shareholders, and, consequently, making LSC committees passive in the management of collective resources. NH is an administrative village in Nanhai district with a total area of 201.05 ha in 2012 of which 58.6% had been leased out for industries. Because of the concentration of industrial enterprises, movement of village members was frequent, but the inflow of population outnumbered the outflow. From 2000 to 2012, the population change affected 38.4% of total members and the new population increase was 11.3%. High volumes of LSC shares were transacted (selling and buying back) at discretionary rates without market price, which was only twice the yearly dividend. This situation was clearly not favorable for sellers, but tremendously beneficial for buyers. The prominent free-rider and short-horizon problem exists for both the aging and young members who do not have incentives to contribute to the long-term opportunity.

As one of the 11 LSCs in NH village, SB LSC had 535 villagers, 180 households, and a total land area of 18.8ha in 2012. Of total land, 61.7% was leased out for industrial use (the remaining as residential land). Land leasing income accounted for 91.0% of the total LSC revenue, and 85.3% of the total LSC revenues were distributed as shareholder dividends. The LSC dividends (52,962 Yuan per household) contributed 35.5% to the annual household income, and rental income from family-house leasing accounted for 15.7%. Nevertheless, according to the authors' questionnaire survey to 92 villagers, in spite of a great deal of dividends distributed, those who were very satisfied with the amount of LSC dividends received were only 6.4%; 37.2% believed that annual dividends could be more; 50% of the interviewed were unsatisfied and the remaining 6.4% were very unsatisfied with the amount of dividends received. Although 54.3% were worried about the long-term sustainability of the land leasing economy, only 15.5% of those interviewed were willing to receive lower dividends in order for the LSC to retain more for improvement and investment. In general, it seems that the majority of LSC members were only interested in the extraction of land rents.

(3)   Underproductive utilization of scarce land resources in LSCs: its ossification

The impacts of institutional land-use change, either formal or informal, unfold along the following three dimensions: first, as a derivative product of officially recognized formal institutional change of the LSC system, LSCs' land conversion and leasing for non-agricultural uses is an informal institution only agreed tacitly by local government. This informal institution leads to an uncertain institutional environment in which LSCs land leasing tends to be makeshift and a hasty action in order to capture land rents in the face of governments' land acquisition. This short-term behavior that will hinder necessary investment in land and land-use efficiency tends to be suboptimal.

Second, given the ambiguously defined rights of LSC members in regard to collective assets (with land rent as the major asset), LSCs suffer from a series of problems, such as free-rider, horizon, and organizational control problems. Members do not have confidence in the long-term future of their LSCs, and prefer short-term returns to long-term economic productivity. The primary concern of village managers tends to be the maximization of short-term revenues for distribution, and they also tend to be risk-averse, and conservative and passive in land management.

It has been shown that in Nanhai district, the natural-village-based LSC only retained 6.4% of its total revenue on average in both 2007 and 2008 for future investment (NHDRA, 2006–2012). In order to avoid capital investment, many LSCs have adopted policy of direct leasing of raw land without services and premises. In *Guicheng* town in Nanhai, for example, of the total industrial land of 1295.8 ha managed by 114 LSCs in 2007, 977.3ha (75.4%) of the land was leased directly to industrial tenants without necessary infrastructure, despite the cooperatives knowing that land rents of premises leasing (1.2million Yuan/ha) could be as much as six times of that of raw land leasing (NHDRA, 2008). The LSCs, as a welfare organization, focus on extracting short-term land rents instead of investing in long-term village economies, further resulting in low-efficient use of scarce land resources.

Third, self-contained development of small-area villages generates, in aggregate, compartmentalized industrialization and fragmented urbanization (Zhu & Guo, 2015). Such a mode of village-based industrialization impedes the economic provision of industrial infrastructure (such as waste water treatment and other necessary services) which is essential for productive manufacturing. Factories are built at a low cost and with simple structures. Substandard industrial estates do not appeal to high-standard industrial tenants, and only those opportunistic companies who do not care about low-standard factories are attracted. For example, in 2012, NH village hosted 469 small industrial plants, with only three plants having an output annual value more than 5 million Yuan (a quality yardstick used by local official statistics).

Given the three causes or dimensions, land in peri-urban Nanhai is underproductive, which is clearly shown through a comparison between two townships. In general, because of equipped infrastructure and facilities, the use of state-owned land, and the existence of an agglomeration economy, urban industrial zones will be more productive than the rural LSC industrial sites. Shishan is a township in Nanhai where the Nanhai Economic Development Zone (NHEDZ) and the largest state-owned urban industrial zone (2,312 ha) is located. In order to make this area a high-quality industrial zone, only industrial plants with an annual output value of more than 5 million Yuan are admitted. Since the inauguration of NHEDZ, which accounts for 25.6% of the total industrial land stock in Shishan, Shishan's industrial land productivity has been enhanced significantly. This is shown by its comparison with Luocun Township where 94.4% of industrial land is administered by LSCs (Table 5.3). Shishan's industrial land productivity was only 60% of Luocun's in 1996. With NHEDZ, Shishan's land productivity was 65% more than that of Luocun in 2008. This demonstrates that integrated urban industrial zones are more productive than fragmented rural industrial patches.

LSC shareholders are averse to investment for long-term productivity, as it would reduce the amount of annual dividends received. Progressive extraction of land rents without adequate investment has LSCs locked in a state of gradual deterioration. The scarce land is used unproductively as industrial land and tenant plants are of low quality. Shareholders pressing for ever-increasing dividends drive LSCs to convert additional farmland to industrial land for leasing until there is no vacant land left. LSC village

Table 5.3 Productivity of industrial land in Shishan and Luocun, 1996–2008

Township	Industrial Output (¥ million)	Industrial Land (ha)	Productivity of Industrial Land (¥ million/ha)
1996			
Luocun	1,904	741	2.57
Shishan	4,583	2,974	1.54
2008			
Luocun	14,155	2,196	6.45
Shishan	96,646	9,035	10.69

Sources:*NHBDPS, 1998–2011;Archives of NHBLR, 2008

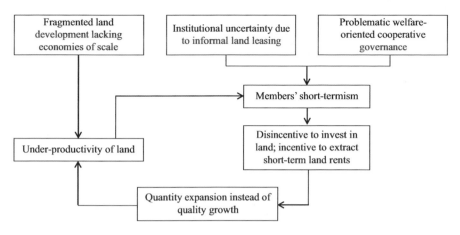

Figure 5.6 The cycle of underproductivity of LSCs' land leasing economies
Source: drawn by the authors

economies are basically of quantitative expansion, instead of quality growth. Without external intervention, village land economies will become stagnant and fixed in the vicious cycle of underproductivity as shown in Figure 5.6.

5.4 Urbanization and land development led by municipal coordination: a case of Kunshan in YRD

As a county-level city under the jurisdiction of the Suzhou Municipality, Kunshan is located to the southeast of Suzhou and to the northwest of Shanghai, sandwiched between these two prosperous cities (Figure 5.7). Before 1989, Kunshan was a rural county, but it has witnessed rapid urbanization and economic growth ever since. It has been ranked as the top city in terms of GDP among counties and county-level cities in China for about 13 consecutive years.

From 1970 to 2015, Kunshan's GDP rose from 100 million Yuan to308 billion Yuan (at current prices). The significant economic growth was greatly driven by the rapid development of manufacturing sector. The share of industrial output of total GDP was 55.1% in 2015, while the proportion of industrial output was only 0.1% in 1970. The economic boom has attracted a large influx of people. In 2015, the number of migrants was up to 1,272,000, or 1.6 times of local population, which was only 787,700 (KSBDPS, 1978–2015). Rapid industrialization has promoted the advance of urbanization. In 2010, 73.8% of the locals were residents with urban *hukou*, reaching 7.3 times of the proportion in 1970, while the remaining

Figure 5.7 Location of Kunshan in Yangtze River Delta and China.
Source: drawn by the authors

locals had rural *hukou*. Construction land also expanded at a fast pace. The area of land for urban settlements and other non-agricultural uses in 2000 (with the exception of rural settlements) was 14 and 8 times of that in 1987. The ratio of construction land area to farmland area changed from 3.2%:96.8% to 28.3%:72.7% from 1970 to 2010 (Zhu, 2017).

5.4.1 Transition from bottom-up to top-down land institutions during rapid rural industrialization

Since the early 1980s, Kunshan has undergone a dynamic rural industrialization under a transitional governance. The explosion of TVEs in the 1980s spurred rural non-agricultural development at a tremendous pace. With the increased industrial expansion along with strong government intervention in the 1990s, concentrated industrial zones gradually replaced the scattered TVEs. Since the early 2000s, villagers have demanded a larger share of land rents during urbanization and industrialization, giving rise to the emergence of LSCs. After the decay of TVEs, land development in Kunshan was dominated by urban governments. Generally speaking, during the rapid rural industrialization, the nature of land institutions in Kunshan was transformed from bottom-up to top-down.

(1) Rural reform and the blossom of TVEs

From the late 1950s to the early 1970s, only communes and brigades were permitted to establish enterprises to produce industrial goods, and these enterprises were only to serve in agricultural production and to meet the demand of daily necessities for peasants. The production teams and individuals were forbidden to run any enterprises so that they could focus only on agricultural production. In this context, Kunshan was mainly engaged in agricultural production during this period, with some small-scale TVEs scattered throughout the countryside. However, the small amount of farmland holding per capita gave rise to the problem of rural surplus labor. To provide more employment opportunities and raise farmers' incomes, rural reform was carried out in the early 1980s. The reform gradually relaxed the constraint over rural non-agricultural activities, releasing a tremendous amount of productive energy in rural areas that had previously been restricted by tight government control. Rural enterprises thrived, especially the TVEs which employed local surplus labor forces, land, and capital, to capture the real or expected opportunities which emerged with China's reform and openness.

With the village leaders' efforts to promote rural industrialization by bringing in expertise (Marton, 2000), 335 TVEs had been established in Kunshan by 1985, accounting for 67.7% of the total 495 industrial firms. The share of industrial output value contributed by the TVEs of the total rose from 59.5% in 1985 to 70.6%in 1990 and further to 74.9% in 1995 (Zhu, 2017). In terms of employment structure, local peasants gradually engaged in non-agricultural jobs. In 1985, the rural labor force numbered 301,869, of which 63.26% worked in the field. However, by 1998, the rural labor force had declined to 247,265 and only 33.01% of them continued to farm(KSBDPS, 1978–2015).

(2) Decay of TVEs and concentrated industrial zones development led by urban states

With the radical reforms towards globalization and marketization in the 1990s, TVEs' inherent weaknesses in enterprise governance and technological inferiority made it difficult for them to compete with foreign and private enterprises (Wei, 2010; Zhu, 2018). Privately owned enterprises and joint ventures gradually took over, while TVEs subsequently either went bankrupt or were privatized in the late 1990s. From 1985 to 2000, the ratio of TVEs' industrial output to total output also sharply decreased from 63.5% to 7.9% (Table 5.4). The share of TVEs as a percentage of total industrial output fell to 0.2% in 2010 and further to 0.05% in 2015.

Kunshan's industrialization in the early 1990s occurred without regional and top-down coordination. Various administrative levels such as city-level

Table 5.4 Industrial output and the share of collective industries in the total in Kunshan

Year	Total Industrial Output	Village-owned Industries	Township-owned Industries	Village-owned Industries a% of the Total	Township-owned Industries as % of the Total
	¥ million	¥ million	¥ million	%	%
1985	1,289.2	325.0	441.8	25.2	34.3
1990	5,829.2	1,272.6	2,848.7	21.8	48.9
1995	18,042.2	2,403.3	9,065.4	13.3	50.2
2000	43,060.8	3,392.3		7.9	
2005	163,331.0	1,568.3		1.0	
2010	580,322.1	1,410.9		0.2	
2015	900,028.3	466.2		0.05	

Source: Kunshan Statistic Yearbook, 1985–2015

government, towns, and villages spontaneously and independently developed their own industrial zones. A red flag industrial area in Bacheng town and a Special Economic and Technological Development Zone (SETDZ) just to the east of Yushan town were single urban units under the administration of Kunshan city-level bureaucracies. Meanwhile, motivated by the desire to increase their own revenues, nearly all of the 20 towns and 467 villages at that time had established industrial zones within their perspective administrative jurisdictions (Figure 5.8). The TVEs together with most of the private and small-scale household-based industrial enterprises were spatially dispersed across the villages and towns (Marton, 2000). Such growth of industries without coordination not only led to the development of broadly similar industrial structures across the region, but also created enormous problems related to the provision of infrastructure, duplication, and waste of capital and land (Marton, 2000). These problems were among the factors resulting in the top-down initiation and the expansion of urban-sponsored development.

With the decay of TVEs and increasing inflow of inward investment, top-down intentions to create concentrated industrial development zones to achieve industrial agglomeration and development gradually emerged in Kunshan after1998. As soon as the TVEs withered, urban government took over the roles of leading industrialization by taking control of industrial land supply and land-use planning. Urban government-dominated industrial parks such as the SETDZ and the town-centered industrial development became two main forms of integrated industrialization. Such spatial integration occurred through the top-down adjustment and amalgamation of the administrative divisions.

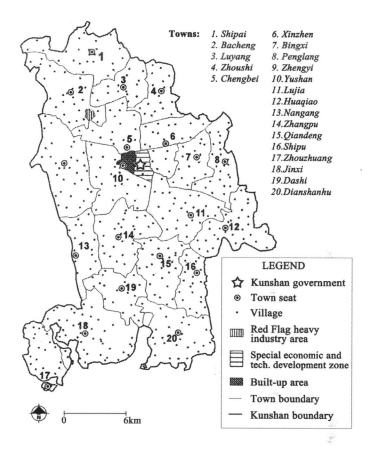

Towns:
1. Shipai
2. Bacheng
3. Luyang
4. Zhoushi
5. Chengbei
6. Xinzhen
7. Bingxi
8. Penglang
9. Zhengyi
10. Yushan
11. Lujia
12. Huaqiao
13. Nangang
14. Zhangpu
15. Qiandeng
16. Shipu
17. Zhouzhuang
18. Jinxi
19. Dashi
20. Dianshanhu

LEGEND
☆ Kunshan government
⊙ Town seat
· Village
▥ Red Flag heavy industry area
▤ Special economic and tech. development zone
▨ Built-up area
— Town boundary
— Kunshan boundary

0 ———— 6km

Figure 5.8 Kunshan administrative divisions, 1998.
Source: drawn by the authors

(3) "Project to enrich people": government-supported collective land and property leasing

In the government-led industrial zones, the land conveyance fees gained by urban governments were high while the amounts of compensation to the villages were minimal. For instance, in 1996, the land conveyance fee in Lujia Township was up to 200,000 Yuan per mu, while the compensation was merely 20,000 Yuan per mu (Qian et al., 2010). The gap between the two fees greatly dampened the enthusiasm of villagers. To solve this problem, the "project to enrich people (fuming project)" policy was formally launched in Kunshan in 2003. This policy allowed villagers to take part in land development and share in the rising land values attendant upon industrialization and urbanization. As mentioned in Section 5.1.3, each village was permitted to convert certain amount of farmland for non-agricultural uses after the cultivation of an equal amount of fallow land, and returned part of land when village land was requisitioned by the government (Po, 2008).

The initiation of the "fuming project" was associated with and facilitated by the establishment of LSCs. Households in a village could voluntarily form a cooperative. Each household would invest a different amount of money within a pre-determined interval into the cooperative and obtain a certain number of shares corresponding to the investment. Then, the cooperative would use the funds to lease some plots of developable land from the village to build properties for rent. At the end of each year, the shareholders would be allocated dividends according to their respective shares. According to Po (2008), more than 80% of the net income of the cooperative was generally allocated as dividends.

By the end of 2010, there were 133 cooperatives established at the village level in Kunshan. The cooperatives covered 19,813 households, which accounted for 17.96% of the total number of rural households in Kunshan (KSBDPS, 1978–2015). The investment of shareholders totaled 418.63 million Yuan and 900.77 million square meters of property space was constructed by the end of 2010. The revenue for allocation reached 103.84 million Yuan. On average, a household would receive 5,241 Yuan. The annual dividend each household received averaged 10% or more of the value of their shares (Po, 2008). On average, one household had 3.54 persons, indicating that the per capita dividend was 1,480 Yuan. Adding the direct allocation of village collectives, the income a person could receive from his/her village collective and the cooperative accounted for nearly 10% of his/her total annual income. Thus, in addition to land acquisition and industrial zone development dominated by urban states as the main means of land development, land leasing by the rural villages and the property development by the LSCs were also important driving forces underlying Kunshan's rapid urbanization.

5.4.2 Municipal coordination of land development: concentrated industrial zone development led by urban states

(1) Manufacturing structure advancement and industrial zone development

Since 1998, Kunshan's manufacturing structure has gone through drastic changes. Facing the challenges brought by economic globalization, low-value-added labor-intensive sectors run by the village enterprises lost their competiveness and were gradually phased out. Meanwhile the high-value-added technology-intensive sectors expanded as inward private enterprises and foreign investment flooded into Kunshan. Inward private firms and joint enterprises showed a preference for the industrial zones where quality infrastructures and facilities were well developed (Zhu, 2017).

To promote the industry development, a large amount of land was acquired by urban states from the rural area in order to establish concentrated industrial zones. From 1983 to 2011, farmland expropriated by urban states accounted for 60% of the total area in a township of Kunshan, and made up 73% of the amounts of land in 22 administrative villages (Zhu, 2013b). The Special Economic and Technological Development Zone (SETDZ) was founded covertly by the Kunshan county government in 1985 and nearly all of the 20townships in Kunshan developed their own town industrial zones in the mid-1990s.

SETDZ was located in the urban downtown core of Kunshan. When it was initially established, the area was only 3.75km^2, increasing to 6.18 km^2 in 1991. In mid-1993, the area formally sanctioned by the central government state council was 10.38 km^2. Because the SETDZ was under the autonomous administration of the Kunshan municipal government, its administrative jurisdiction extended east and south into and even superseding some towns. The administrative area had sharply expanded from 20km^2 in 1995 to 57 km^2 in 2004. By the end of 2005, the area reached 93 km^{2}and continued expanding. The concentrated industrial development and deliberate strategy of functional specialization to develop some targeted industrial sectors such as IT industries, high-precision machinery, and refined chemicals attracted a large amount of foreign investment. At the end of 1992, the area accommodated 150 industrial enterprises among which 106 received some form of foreign investment. By 2011, the industrial output of SETDZ accounted for 60.3% of the total in Kunshan, while village industries only made up 0.2% (Zhu, 2017).

(2) Administrative adjustment and integration of industrial development zones

Despite some large-scale industrial zones like SETDZ, there were many industrial zones since each town had at least one industrial zone (Marton, 2000). The administrative adjustment to integrate the original relatively dispersed and separate industrial development was also applied to the towns. Restructuring of township jurisdiction was carried out in 1990 when 22 townships were consolidated to 20, and to 15 in 2001 and further to just 10 in 2005 (KSBDPS, 1978–2015).

As a result, the town administrative jurisdictions were greatly enlarged, making coordination at a larger spatial scale possible. This also facilitated the development of town-centered industrial parks. As shown in Figure 5.9, industrial land existed in each of the former 20 towns in 2008, surrounding the town centers. However, in each of the amalgamated towns, there was an expanding and dominant industrial park. According to Zhu's analysis (2017), the consolidation of townships facilitated industrial agglomeration remarkably. It is noted that the jurisdiction of three townships in the south (which is more rural in nature in comparison to the rest of the county) remained unchanged over the same period, and drastic restructuring

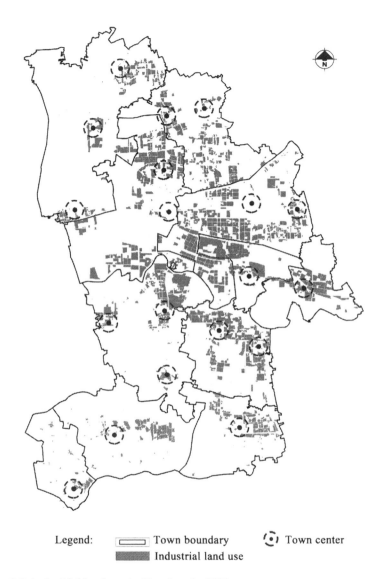

Legend: ▭ Town boundary   ⊙ Town center
           ▓ Industrial land use

Figure 5.9 Industrial land use in Kunshan in 2008
Source: Kunshan Urban Planning Bureau

occurred to those urbanizing townships around the county central area
(Figure 5.10). As a result, nearly all of the restructured townships witnessed
a conspicuous increasing share of industrial output when compared with
total municipal industrial output, while the industrial output percentage of
the unchanged townships in the south declined.

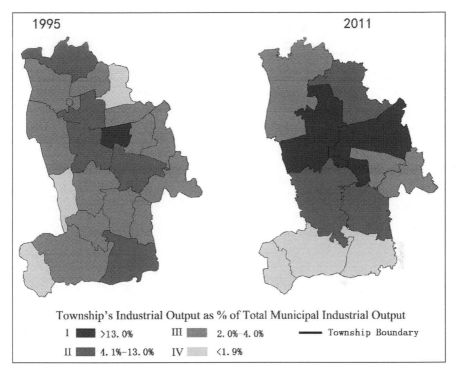

Figure 5.10 Industrial agglomeration from rural factories to industrial zones.
Source: Zhu, J. M. (2017)

(3) Municipal coordinated land rents distribution and social equity

During this period, a substantial amount of land rents were generated during the conversion of rural collective land to state-owned land. Land conveyance fees, or land rent, which came from urban land leasing and were mostly taken by the local municipal government at its disposal, have increasingly become an important financial source for public investment in urban facilities and infrastructure (Zhu, 2016). In contrast to Nanhai where land rents were shared as welfare within the LSCs, Kunshan's land rents were captured by villages as well as governments at all levels, and were mainly used for public investment. From 2001 to 2003, the land rents obtained by the Kunshan municipal government accounted for 68.0%–78.5% of the total (Table 5.5). From 2001 to 2004, 70.8% of the land rent differential gained though land development for industrial use was spent on land consolidation and infrastructure construction (Wang et al., 2006).

Moreover, a municipal-level coordination in Kunshan, to a great extent, mitigated the impact of uneven land rents distribution among villages.

Table 5.5 Land rents distribution during land acquisition (1,000 Yuan per mu, %)

Year	Central Government		Provincial Government		Suzhou Municipal Government		Kunshan Municipal Government		Villages and Village Collectives	
	I	P	I	P	I	P	I	P	I	P
2001	0.28	3.58	0.3	3.81	0.11	1.37	5.38	68.08	1.83	23.16
2002	0.27	2.35	0.3	2.59	0.11	0.93	9.13	78.57	1.81	15.57
2003	0.32	2.62	0.31	2.55	0.11	0.87	9.47	77.14	2.07	16.82

Source: Liu, F., Qian, Z. H., & Guo, Z. X. (2006)
Note: I represents the land rent differential per mu; P represents the proportion of land rent differential.

A fiscal transfer policy was adopted by the Kunshan municipal government to support rural townships' development in the name of farmland conservation and ecological protection. Sharing urbanization welfare in the form of a fiscal transfer to rural villages seemed just and fair. In 2012, the amount of fiscal transfers made up34.7% of the total annual village collective revenues. The fiscal transfers accounted for as much as 56.1% of collective revenues for the villages in Jingxi, but only 8.5% for the villages in Huaqiao (Zhu, 2017). After municipal fiscal transfers to the villages, the minimum and mean levels of village collective income increased significantly. Generally speaking, municipal coordination of land rents distribution not only promoted sustainable development by improving infrastructure and facilities, but also helped achieve social equality through fiscal transfer policy.

5.4.3 Industrial structure upgrading, spatial integration, and better use of land

(1)   Sustainable economic development and industrial structure upgrading

Over the past four decades, Kunshan has experienced great economic growth. From 1980 to 2015, its GDP increased from 300 million Yuan to 308 billion Yuan, with an annual growth rate of 21.9% (KSBDPS, 1978–2015). The temporal change in the sector distribution of GDP explicitly shows that Kunshan's economic miracle was mainly driven by manufacturing and implicitly illustrates that there were different driving forces underpinning the GDP growth for different time periods. During the 1970s when Kunshan was a complete rural county, the ratio of GDP in manufacturing was below 30% and agriculture was the dominant sector contributing to GDP (Figure 5.11). Immediately after China's rural reform in 1980, the

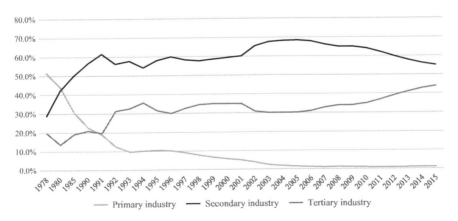

Figure 5.11 Change in the industrial structure over time, Kunshan, 1978–2015.
Source: KSBDPS (1978–2015)

share of manufacturing as GDP increased to 42%. After constant growth in the 1980s, the share of manufacturing as GDP grew to near 60% and remained stable throughout the 1990s, and this growth coincided with the TVEs-led industrialization during this period. Since the administrative adjustment in 2001 to advocate for integrated and large-scale industrial development, the ratio of manufacturing in GDP increased to above 60%. This implies that urban-sponsored industrial development advanced the economic growth during this period.

The urban-sponsored integrated industrial development greatly promoted the upgrading of industrial structure, which is shown in Figure 5.12. The manufacturing industries could often be categorized into three sectors according to the intensity of utilization of various production elements: the tech-intensive sector, the capital-intensive sector, and the labor-intensive sector.[2] In general, the labor productivity of the tech-intensive sector was the highest among the three. The next was capital-intensive sector, and the labor-intensive sector had the lowest labor productivity.

In Kunshan, the averages of industrial output per labor between 2006 and 2015 for the above three sectors were 982,900, 709,800, and 534,700 Yuan, respectively (KSBDPS, 1978–2015). Before 2003, Kunshan's industries were mainly labor-intensive, which accounted for more than 40% of the total output of industrial enterprises which had an annual industrial output of more than 500 million Yuan. However, since 2001, the development of tech-intensive industries has increased, driven by urban governments' active policy to attract such industries. After only two years, the situation reversed, with tech-intensive industries becoming the dominant part of the city industry.

Over time, the industrial structure had been upgraded, which was in sharp contrast with Nanhai where urbanization was mainly dominated by rural villages in terms of self-contained land development. In Nanhai, the industrial structure upgrading was slow or stagnant, which is shown in Figure 5.12. The tech-intensive and capital-intensive industries in Nanhai increased only slightly between 2007 and 2008. The averages for industrial output per labor between 2006 and 2015 were 705,700, 708,200, and 728,200 for the three respective sectors (NHDRA, 2006–2012). The technique- and capital-intensive sectors in Kunshan had higher productivity than those in Nanhai. The productivity of the labor-intensive sector in Nanhai was slightly higher than that in Kunshan. The comparison in industrial structure change between Kunshan and Nanhai shows that concentrated industrial development dominated by the urban government is more efficient than dispersed rural industrialization.

(2) Spatial integration and better use of land

With the restructuring of industry, factories were agglomerated into the industrial zones gradually. Correspondingly, the industrial landscape evolved from scattered village factories to concentrated city and town industrial

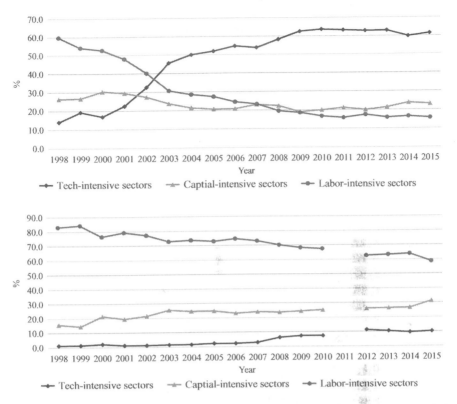

Figure 5.12 Industrial structure change by sectors in Kunshan (upper) and Nanhai (low), 1998–2010.

Source: KSBDPS, NHBDPS (1998–2011)

zones. In this section, the patch analysis in the discipline of landscape ecology is employed to analyze Kunshan's landscape change over time.

Generally speaking, Kunshan experienced a process of gradual frag-mented non-agricultural landscape followed by a tendency of land-use inte-gration. This process is clearly shown in Table 5.6. In 1989, construction land only accounted for 3.19% of the total territory of Kunshan (92,633 ha). The ratio increased to more than 11% by 2000. Meanwhile, the patch numbers of both construction land and farmland rose, so did the density. Between 2000 and 2005, despite that the share of construction land area and the average patch size almost doubled, the number of construction land patches and patch density declined slightly, while farmland patches increased sharply. In the later period to 2010, construction land area further increased, and the numbers of both farmland and construction land patches decreased significantly.

Table 5.6 Patch indices for farmland and construction land in Kunshan

Year	Landscapes	Percent of Landscape (%)	Patch Number	Patch Density	Mean Patch Size (ha)
1989	Farmland	96.81	195	0.21	460.04
	Construction land	3.19	598	0.64	4.95
1995	Farmland	88.77	494	0.53	166.53
	Construction land	11.23	791	0.85	13.15
2000	Farmland	88.29	510	0.55	160.42
	Construction land	11.71	1,384	1.49	7.84
2005	Farmland	74.51	2,943	3.18	23.46
	Construction land	25.49	1,370	1.48	17.24
2010	Farmland	71.69	1,711	1.85	38.83
	Construction land	28.31	795	0.86	32.99

Source: The estimations are based on data from satellite remote sensing maps at http://datamir ror.csdb.cn/admin/introLandsat.jsp

The process is consistent with Kunshan's urbanization process which was driven by rural enterprise first and later by the urban-sponsored development of industrial parks. As mentioned above, urbanizationfrom1989 to 2000 was driven by rural enterprises. Rural industrialization led to the scattering of industrial land uses across the villages and formed a landscape in which residential, industrial, and agricultural land uses densely mixed in with that of villages (McGee et al., 2007). The decrease of construction land patch density between 2000 and 2005 exhibits the preliminary effect of land-use integration and administrative amalgamation implemented in the early 2000s.

The farmland fragmentation during that period was brought about by the extensive construction of roads connecting towns and villages and the increase of the number of farmland plots temporarily left inside the built-up area. The sharp decline of farmland and construction land numbers shows the great positive effect in advancing integrated land development of the amalgamation of small towns after 2005 and the expansion of SETDZ from the center outward to spatially incorporate the adjacent town industrial parks. As revealed by the analysis, the spatial agglomeration of industries heavily contributed to the enhancement of environmental integrity and the compact development of Kunshan, which to some extent has helped achieve sustainable urbanization.

Additionally, land-use efficiency was significantly improved due to the agglomeration of factories and the advancement of industry. In 1978, the industrial output value per km^2 was 970,200 Yuan, and it increased to 23.7 million Yuan in 1994. By 2009, the industrial output value per km^2 was 24.25 times and 368.71 times of those in 1978 and 1994, respectively (Chen, 2011).

5.5 Summary

Urbanization patterns are shaped by modes of land development, or more precisely, by a variety of interactions among stakeholders in land markets and development processes under certain institutional settings. Peri-urbanization in China, as has been depicted above, is mainly a result of interactions between rural actors (village collective organizations and individual villagers) and local governments under a setting of ever-changing rural land institutions. The change of rural land institutions, in the spirit of China's gradual reform, has occurred not by design, but through competition and compromise among the stakeholders. As a result, change in rural land institutions has been gradual, incremental, and path-dependent.

Based on a general trend of rural land institutions moving towards strict control of land conversion for non-agricultural uses, clarification of collectively owned rural land, and encouragement of land circulation, there are at least two specific characteristics of the change: 1) the gradual rural land institutional change leads to transitional institutions which are tentative in nature and result in uncertainty of institutional settings. The uncertainty may either induce disorderly competition and short-term behavior in emerging land markets and diverse social fabrics or promote further institutional invention guided by the government to avoid market failure (Schotter, 1981; Zhu, 2005); 2) Decentralization as an instrument to initiate economic reform devolves the decision-making power from the central state to the local governments, which leads to a variety of bottom-up institutional innovations across localities. It is under such a setting that the LSC system arises and differs in terms of land rights assignments in different regions. The various modes of interactions among stakeholders under local-specific land institutions lead to different patterns of land development and peri-urbanization.

Nanhai and Kunshan are employed as two representative cases to elaborate on the above statements. For both cities, the rapid development of TVEs spearheaded economic development in their early stages. However, different ways of land rents distribution among the governments and rural actors after the withdrawal of TVEs allowed the two cities to take different trajectories in land institution assignment, land development modes, urbanization patterns measured by urban forms, land-use efficiency, and even economic sustainability. In Nanhai, rural collectives strengthened their land rights relative to the local government through bottom-up establishment of LSCs which were village-based and welfare-oriented. The rural collectives in Nanhai competed for land rents with local governments in their emerging and informal land leasing markets.

Self-contained land development on a small scale leads to extremely fragmented urban forms, causing land to be constrained in a circle of unproductive use. Further institutional change dependent on the path is needed to make land use efficient and economic development sustainable in Nanhai. In contrast, land development in Kunshan is coordinated by local governments

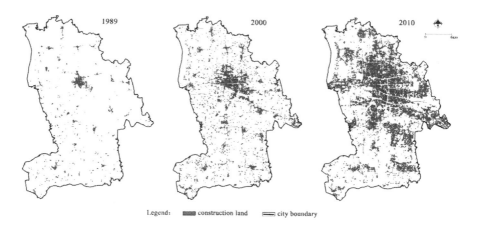

Figure 5.13 Landscape changes over time, 1989, 2000, 2010.
Source: Remote sensing maps of Kunshan from the website of www.gscloud.cn/

at larger scales and in an orderly manner. The LSCs are investment-oriented. Land use tends to be integrated and is much more efficient (Figure 5.13).

Notes

1 In LML, it is regulated that starting TVEs and establishing joint-ventures with other enterprises can convert agricultural land after the approval of the urban government (Article 60, LML, China 1999). Only recently has agricultural land leasing been allowed to overcome the problem of farmland fragmentation. Spontaneous land conversion for non-agricultural uses is still under the strict supervision of the urban governments in order to protect the scarce farmland, and land leasing for non-agricultural uses is not allowed, or under strict control.

2 The tech-intensive sector includes industries like the electronic and telecommunication equipment manufacturing industry and instrument, meters, cultural and office equipment industry. The capital-intensive sector consists of the manufacturing industries like general machine, special equipment, transportation equipment, electrical machine and equipment, the pharmaceutical industry, and the chemical manufacturing industry. The labor-intensive industries are like textile industry, food processing and manufacturing industry, etc.

References

Alpermann, B. (2001). The Post-Election Administration of Chinese Villages. *The China Journal*, 46:45–67.

Barzel, Y. (1989). *Economic Analysis of Property Rights*. Cambridge: Cambridge University Press.

Brandt, L., Rozelle, S., & Turner, M. A. (2002). Local Government Behavior and Property Right Formation in Rural China. *Journal of Institutional and Theoretical Economics-Zeitschrift Fur Die Gesamte Staatswissenschaft*, 160(4):627–662.

Brown, L. R. (1995). *Who Will Feed China?: Wake-Up Call for a Small Planet.* New York: W W Norton & Company.

Byrd, W. A., & Lin, Q. (1990). *China's Rural Industry: Structure, Development, and Reform.* New York: Oxford University Press.

Che, J., & Qian, Y. Y. (1998). Insecure Property Rights and Government Ownership of Firms. *Quarterly Journal of Economics,* 113(2):467–496.

Chen, H. Y. (2011). *Study on Mechanism of Land Intensive Use Which Based on Economic Development Pattern Transformation – A Case of Kunshan County in Jiangsu Province.* Doctoral dissertation, Nanjing Agricultural University, 91. (In Chinese).

Chen, J. (2006). Collective Ownership of Rural Land: Tenure or Principal-Agent Problem?*Economic Research,* 7:83–91. (In Chinese).

Cook, M. L.(1995). The Future of US Agricultural Cooperatives: A Neo-Institutional Approach. *American Journal of Agricultural Economics,* 77(5):1153–1159.

Fu, C. (2003). *Nongcun shequxing gufenhezuozhi yanjiu (Research of the Rural Community-Based Shareholding Co-Operatives).* Beijing: Zhongguo jingji chubanshe (China Economy Press). (In Chinese).

Fu, C., & Davis, J. (1998). Land Reform in Rural China since the Mid-1980s. *Land Reform,* 1998(2):122–137.

Gao, M. C. F. (1999). *Gao Village: A Portrait of Rural Life in Modern China.* Honolulu: University of Hawaii Press.

Guo, X. L. (2001). Land Expropriation and Rural Conflicts in China. *The China Quarterly,* 2001(166):422–439.

Ho, P. (2001). Who Owns China's Land? Property Rights and Deliberate Institutional Ambiguity. *The China Quarterly,* 2001(166):394–421.

Ho, P., & Lin, G. C. S. (2003). Emerging Land Markets in Rural and Urban China: Policies and Practices. *The China Quarterly,* 175(175):681–707.

Hou, W. Y., & Zeng, J. M. (1991). A Survey Report of Agricultural Workers' Employment Structure. *High Education Exploration,* 2:34–38. (In Chinese).

Hsing, Y. T. (2006). Brokering Power and Property in China's Townships. *The Pacific Review,* 19(1):103–1214.

Huang, Y. (1996). *Inflation and Investment Control in China: The Political Economy of Central-Local Relations during the Reform Era.* Cambridge: Cambridge University Press.

Krug, B., Ed. (2004). *Advances in Cooperative Theory since 1990: A Review of Agricultural Economics Literature.* Restructuring Agricultural Cooperatives. Amsterdam: Erasmus University Rotterdam.

KSBDPS (Kunshan Bureau of Development Planning and Statistics). (1978–2015). *Kunshan's Statistical Yearbook.* Suzhou: Kunshan Press. (In Chinese).

Kung, J. K. (1995). Equal Entitlement versus Tenure Security under a Regime of Collective Property Rights: Peasants Preference for Institutions in Post-Reform Chinese Agriculture. *Journal of Comparative Economics,* 21(1):82–111.

Kung, J. K. S. (2000). Common Property Rights and Land Reallocations in Rural China: Evidence from a Village Survey. *World Development,* 28(4):701–719.

Land Management Law. (1988). www.china.com.cn/law/flfg/txt/2006-08/08/content_7063905.htm, accessed on October 19th, 2018. (In Chinese).

Land Management Law. (1999).www.gov.cn/banshi/2005-05/26/content_989.htm, accessed on October 19th, 2018. (In Chinese).

Lau, L. J., Qian, Y. Y., & Roland, G. (2000). Reform without Losers: An Interpretation of China's Dual-Track Approach to Transition. *Journal of Political Economy*, 108(1):120–143.

Lin, G. C. S., & Ho, S. P. S. (2005). The State, Land System, and Land Development Processes in Contemporary China. *Annals of the Association of American Geographers*, 95(2):411–436.

Lin, J. Y. F., & Liu, Z. Q. (2000). Fiscal Decentralization and Economic Growth in China. *Economic Development and Cultural Change*, 49(1):1–21.

Liu, F., Qian, Z. H., & Guo, Z. X. (2006). External Profit, Consistency and Innovation of Fumin Shareholding Cooperative System in Kunshan – Analysis from the Perspective of Institutional Economics. *Agricultural Economic Problems*, 2006(12):54–60. (In Chinese).

Liu, S. Y., Carter, M. R., & Yao, Y. (1998). Dimensions and Diversity of Property Rights in Rural China: Dilemmas on the Road to Further Reform. *World Development*, 26(10):1789–1806.

LML (Land Management Law). (2004). State Council of China.

Marton, A. M. (2000). *China's Special*. London and New York: Routledge.

McGee, T. G., Marton, A., Lin, G. C. S., Wu, J., & Wang, M. (2007). *China's Urban Space: Development under Market Socialism*. London: Routledge.

Ministry of Agriculture, China. (1993). Survey on Agrarian Production under the Household Responsibility System in China. *Issues of Agricultural Economy*, 11:45–52. (In Chinese).

Ministry of Construction of the People's Republic of China. (2008). *The Methods of Housing Registration*. Beijing: Official Document.

Ng, M. K., & Xu, J. (2000). Development Control in Post-Reform China: The Case of Liuhua Lake Park. *Cities*, 17(6):409–418.

NHAB (Nanhai Archives Bureau). (1982). *The Third Nanhai Population Census*. (Unpublished archives, in Chinese).

NHBDPS (Nanhai Bureau of Development Planning and Statistics). (1998–2011). *Nanhai's Statistical Yearbook*. Foshan: Nanhai Press. (In Chinese).

NHDRA(Nanhai Department of Rural Affairs). (2006–2012). *Nanhai's Rural Economic Statistic Yearbook*. Foshan: Nanhai Press. (In Chinese).

North, D. C. (1991). Institutions. *The Journal of Economic Perspectives*, 5(1):97–112.

O'Neill, R. V., Krummel, J. R., Gardner, R. H., Sugihara, G., Jackson, B., DeAngelis, D. L., Milne, B. T., Turner, M. G., Zygmunt, B., Christensen, S. W., Dale, V. H., & Graham, R. L. (1988). Indices of Landscape Pattern. *Landscape Ecology*, 1(3):153–162.

Oi, J. C. (1992). Fiscal Reform and the Economic Foundations of Local State Corporatism in China. *World Politics*, 45(1):99–126.

Oi, J. C. (1995). The Role of the Local State in China's Transitional Economy. *China Quarterly*, 1995(144):1132–1149.

Oi, J. C., & Walder, A. G. (1999). *Property Rights and Economic Reform in China*. Stanford, CA: Stanford University Press.

Ortmann, G. F., & King, R. P. (2007). Agricultural Cooperatives I: History, Theory and Problems. *Agrekon*, 46(1):28.

Park, A., & Shen, M. (2003). Joint Liability Lending and the Rise and Fall of China's Township and Village Enterprises. *Journal of Development Economics*, 71(2):497–531.

Pei, X. (2002). The Contribution of Collective Landownership to China's Economic Transition and Rural Industrialization: A Resource Allocation Model. *Modern China*, 28(3):279–314.

Po, L. (2008). Redefining Rural Collectives in China: Land Conversion and the Emergence of Rural Shareholding Cooperatives. *Urban Studies*, 45(8):1603–1623.

Po, L.(2011). Property Rights Reforms and Changing Grassroots Governance in China's Urban–Rural Peripheries: The Case of Changping District in Beijing. *Urban Studies*, 48(3):509–528.

Porter, P. K., & Scully, G. W. (1987). Economic Efficiency in Cooperatives. *Journal of Law & Economics*, 30(2):489–512.

Qian, Y. Y. (2000). The Process of China's Market Transition (1978–1998): The Evolutionary, Historical, and Comparative Perspectives. *Journal of Institutional and Theoretical Economics-Zeitschrift Fur Die Gesamte Staatswissenschaft*, 156(1):151–171.

Qian, Y. Y., & Weingast, R. (1997). Federalism as a Commitment to Preserving Market Incentives. *Jounal of Economic Perspectives*, 11(4):83–92.

Qian, Z. H., Ji, X. Q., & Liu, F. (2010). External Profit, Consensus and Innovation of Land Use System in Rural Area – A Theoretical Analysis of Fumin Shareholding Cooperative System in Kunshan. *Case Study of China's Institutional Change*, 2010(8):171–192. (In Chinese).

Royer, J. S. (1995). Potential for Cooperative Involvement in Vertical Coordination and Value-Added Activities. *Agribusiness: An International Journal*, 11(5):473–481.

Rozelle, S., & Li, G. (1998). Village Leaders and Land-Rights Formation in China. *American Economic Review*, 88(2):433–438.

Schotter, A. (1981). *The Economic Theory of Social Institutions*. Cambridge: Cambridge University Press.

Staatz, J. M. (1987). Recent Developments in the Theory of Agricultural Cooperation. *Journal of Agricultural Cooperation*, 2(20):74–95.

Tang, J. J. (2009). Sanji Suoyou, Dui wei Jichu (Collective Ownership Belonging to Three Entities (the Commune, the Brigade, and the Team), and with the Team Being the Basic Holder. *Archives World*, 4:17–21.

Tang, Y. (1989). Urban Land Use in China: Policy Issues and Options. *Land Use Policy*, 1989(6):53–63.

Tao, R., Xi, L., Fubing, S., & Hui, W. (2009). China's Transition and Development Model under Evolving Regional Competition Patterns(Diqu jingzheng geju xia de Zhongguo zhuangui: caizheng jili he fazhan moshi fansi). *Economic Research*, 2009(7):21–33. (In Chinese).

Tian, L., & Zhu, J. (2013). Clarification of Collective Land Rights and Its Impact on Non-Agricultural Land Use in the Pearl River Delta of China: A Case of Shunde. *Cities*, 35(4):190–199.

Vitaliano, P. (1983). Cooperative Enterprise: An Alternative Conceptual Basis for Analyzing a Complex Institution. *American Journal of Agricultural Economics*, 65(5):1078–1083.

Wang, H. (1994). *The Gradual Revolution*. New Brunswick, NJ: Transaction

Wang, X. Y., He, M. Y., & Gao, Y. (2006). An Empirical Study on the Distribution of Land Rent Differential in the Conversion of Agricultural Land in China – Analysis Based on the Sample Survey of Kunshan, Tongcheng and Xindu. *Management World*, 2006(5):88–95. (In Chinese).

Wei, Y. D. (2010). Beyond New Regionalism, Beyond Global Production Networks: Remaking the Sunan Model, China. *Environment & Planning C Government & Policy*, 28(1):72–96.

Wen, T. (1998). Sources of Assets of TVEs and Principles in the Reform ofTVEs (Xiangzhen Qiye Zichan de Laiyuan yu Gaizhi zhong de Xiangguan Yuanze). *Zhejiang Soc. Sci.*, 1998(3):38–41. (in Chinese).

Wen, T. J., & Zhu, S. Y. (1996). Zhengfu Ziben Yuanshi Jilei Yu Tudi "Nonzhuanfei" (Governments' Captial Accumulation and Conversion of Farmland to Nonagricultural Land). *Guanli shijie [Management World]*, 5:161–169. (In Chinese).

Wong, C. P. W. (1992). Fiscal Reform and Local Industrialization: The Problematic Sequencing of Reform in Post-Mao China. *Modern China*, 18(2):197–227.

Wong, C. P. W., Heady, C., & Woo, W. T. (1995). *Fiscal Management and Economic Reform in the People's Republic of China*. Hong Kong: Oxford University Press.

World Bank. (2002). *Building Institutions for Markets: World Development Report 2002*. New York: Oxford University Press.

Yao, Y. (2000). An Analytical Framework of the Institutions of Rural Land in China. *China Social Science*, 2:54–65. (In Chinese).

Zhou, Q. R., & Liu, S., Eds. (1988). *Meitan: The Transformation of Land System of a Traditional Agricultural Zone (Meitan: yige chuantong nongqu de tudizhidu bianqian*. China's contemproprary land system (Zhongguo dangdai tudizhidu). Changsha, China: Hunan Science and Technique Press.

Zhu, J. M. (1994). Changing Land Policy and Its Impact on Local Growth: The Experience of the Shenzhen Special Economic Zone, China, in the 1980s. *Urban Studies*, 31(10):1611–1623.

Zhu, J. M. (1999). Local Growth Coalition: The Context and Implications of China's Gradualist Urban Land Reforms. *International Journal of Urban and Regional Research*, 23(3):534–550.

Zhu, J. M. (2005). A Transitional Institution for the Emerging Land Market in Urban China. *Urban Studies*, 42(8):1369–1390.

Zhu, J. M. (2013a). Coordinated Urban-Rural Development: Urban Integrity and Rural Autonomy. *Urban Planning Forum*, 44(1):10–17. (In Chinese).

Zhu, J. M.(2013b). Governance Over Land Development during Rapid Urbanization under Institutional Uncertainty, with Reference to Periurbanization in Guangzhou Metropolitan Region, China. *Environment and Planning C: Government and Policy*, 31(2):257–275.

Zhu, J. M. (2016). The Impact of Land Rent on the Formation of Urban Structure and Urban Renewal during Institutional Change. *Urban Planning Forum*, 2016 (2):28–34. (In Chinese).

Zhu, J. M. (2017). Making Urbanisation Compact and Equal: Integrating Rural Villages into Urban Communities in Kunshan, China. *Urban Studies*, 54(10):2268–2284.

Zhu, J. M. (2018). Transition of Villages during Urbanization as Collective Communities: A Case Study of Kunshan, China. *Cities*, 72:320–328.

Zhu, J. M., & Guo, Y. (2014). Fragmented Peri-Urbanisation Led by Autonomous Village Development under Informal Institution in High-Density Regions: The Case of Nanhai, China. *Urban Studies*, 51(6):1120–1145.

Zhu, J. M.,& Guo, Y. (2015). Rural Development Led by Autonomous Village Land Cooperatives: Its Impact on Sustainable China's Urbanisation in High-Density Regions. *Urban Studies*, 52(8):1395–1413.

Zhu, J. M., & Hu, T. T. (2009). Disordered Land-Rent Competition in China's Periurbanization: Case Study of Beiqijia Township, Beijing. *Environment and Planning A*, 41(7):1629–1646.

6 Institutional change to redevelop peri-urban areas and spatial lock-in

In 2009, the Guangdong provincial government launched their "Three-olds Renewal (TOR)" polices in order to carry out land reform and push forward the redevelopment of inefficient construction land. According to the TOR policies, the government delegated certain powers to existing landholders in order to facilitate the redevelopment of old towns (dilapidated urban areas), old villages (deteriorating villages), and old factories (underused industrial land parcels) and to formalize the informal land rights of rural collectives over collective land in peri-urban areas.

Applying the framework of institutional change, this chapter examines the new arrangement of TOR policies, looking closely at the role of the government and power-relations among stakeholders. By taking the typical peri-urban areas, Nanhai and Panyu in the PRD as cases, this chapter presents local practices in which institutional changes are made in the redistribution of benefits among the government, developers, village collectives, and villagers. However, a spatial lock-in effect due to reliance of rural collectives on landed interests, high transaction costs to achieve consensus for redevelopment, and still ambiguous land rights inside the collectives has made redevelopment of peri-urban areas a time-consuming and complicated process. Specific processes of the TOR policies and reasons for spatial lock-in have also been elaborated upon through detailed case studies of implemented projects.

6.1 TOR: policy and process

6.1.1 Background behind the TOR

As mentioned in Chapter 5, in the PRD region, urbanization has been mainly driven by bottom-up rural industrialization, which has led to a rapid expansion of the built-up area since China's reform and openness. Rural industrialization began in the early 1980s when the TVEs began to develop rapidly. Facing increasingly fierce competition from private sectors, the TVEs gradually withered in the late 1990s. However, it did not slow the speed of rural non-agricultural development. Instead, rural villages sped up their non-agricultural development based on LSCs. The cooperatives re-collectivized the land distributed to individual households in HPRS and managed all of a village's land except for housing sites. In the PRD region, collective construction land has accounted for more than 60% of total construction land in many areas.

(1)   Government's attempts to formalize informal land development

The government was concerned with the prevalent informal land development led by LSCs as early as 2003 when it first attempted to deal with it. In 2003, the Guangdong provincial government issued "Notes of Experimenting the Circulation of Rural Non-agricultural Land," aiming to formalize the land leasing market. This policy stated: 1) Rural construction land is allowed to circulate in terms of leasing, conveyance, sub-leasing, transfer, and mortgage, as long as the uses of plot satisfy land-use planning and urban planning, and landholders obtain certificates of Land Use Rights (LURs) from the government and landownership is clear and without any disputes; 2) The land is allowed for any use except commodity real estate; 3) Circulation of land should obtain approval of two-thirds of villagers; 4) After signing contracts with land tenants, landholders should apply to the land management department of the higher level government for registration of the transaction; 5) Landholders should pay some fees and fines to the government for land circulation. In 2005, the Guangdong provincial government further issued "Regulations of the Circulation of Land Uses Rights of Collectively owned Construction land," which is called the "Land New Deal" on the ground that this regulation officially recognizes the transfer rights of collective land, although villagers' housing sites and houses are forbidden to circulate.

Generally speaking, as early as the mid-2000s, local governments had sensed the necessity to effectively manage rural land leasing for non-agricultural uses, and attempted to do so through issuing regulations on formalizing LURs and other bundles of rights. However, due to the formation of landed interests in increasing land conversion, high costs of formalizing extant land leasing, and an unclear attitude of the governments towards increasing land conversion (permission or prohibition), the regulations did not generate effects as expected.

(2)   Transition from Greenfield development to redevelopment

Since the mid-2000s, especially after the 2008 global financial crisis, China has entered into a new round of economic, social, and spatial transformation. After nearly two decades of relying on low-end industries, the country began to realize the significance of upgrading traditional industries, and advocated transforming the economy from being manufacturing-driven to services-oriented. With economic transformation, urban policy has put more emphasis on development rather than growth.

In order to promote urban transformation, the central government introduced strict control over urban expansion through a land quota system in 2003. In the mid-2000s, many cities in the PRD region had already converted more than half of their territory to construction land. Facing the limited developable land, the Guangdong provincial government proposed a policy of "Emptying the Cage to Attract New Birds (tenglong huanniao)," aiming to encourage relocation of low-end industries and leave land for high-tech industries. Under the strict control of the central state over land quota, local governments have had to control the conversion from agricultural land into construction land.

Such strict control over land conversion suddenly stopped the extant ways by which rural collectives maintained the growth of their collective income. It led to the following responses from rural collectives: 1) Some LSCs still carried out the unapproved land conversion as a resistance. As a response, local governments adopted coercive measures of control, such as a suit against village cadres who violated the rules and even arrested those who did not follow the rules; 2) rural villages took collective actions to impede agricultural land expropriation by local government, as the change of land ownership needs agreement of at least two-thirds of shareholders. Alleviation of such a severe confrontation between the government and rural villages required urgent institutional change. Accordingly, further institutional change has had to consider how to provide rural villages with new opportunities of income growth after closing a window. It is just under such a backdrop that the TOR program has come into being.

(3)   Pilot practice and policy release

Foshan City is among the experimental sites in Guangdong province for piloting the TOR program. As early as in 2007, Foshan municipal government issued a series of regulations of guiding the experiment, and it required that every district draft a seven-year redevelopment plan and issue specific regulations based on the city-level policies to guide redevelopment. In 2008, the Nanhai district government initiated TOR. At the end of 2008, Guangdong province signed a contract with the Ministry of Land and Resources to construct a pilot province of promoting more economical and intensive land use. The first contract period lasted from 2009 to 2012. In 2009, based on practices of the abovementioned pilot cities over two years, the Guangdong provincial government issued an official regulation of "Opinions about Carrying out TOR to Promote Economical and Intensive Use of Land." This regulation represented that TOR had been officially agreed upon, and formal institutions came into being.

6.1.2 *Policy design of TOR*

(1)   Clarification of land property rights

As mentioned before, by law, rural collective land belongs to the collective of villagers, but who represents the collective has never been clearly defined.

Although under the LSC system, land has been re-collectivized to be managed by cooperatives, land ownership has never been officially registered. The boundaries between different cooperatives have never been clarified and the governments never issued cooperatives official certificates. The principle of formalization is to respect existing landholding conditions (Lin, 2015). After clarifying locations, boundaries, and areas, the governments defined land parcels to be owned by corresponding cooperatives and issued certificates of land ownership to them.

Only after the clarification of land ownership, can the bundles of rights be redefined. The primary right is redefinition of LURs. The construction land is distinguished between legal and illegal, judged by whether landholders have gone through land conversion approval procedures and received certificates of LURs. For illegally converted land, landholders are permitted to re-apply for various permits after paying any related fees and fines in full. After the payment, the governments issue certificates of LURs to them according to land uses of *status quo*. After receiving both the certificates of land ownership and LURs, villages can then enjoy more bundles of other rights in redevelopment. If villages decide not to change collective land ownership, the land can be circulated through leasing, conveyance, transference, or mortgage for uses other than commodity housing.

Villages can also apply to urban governments for a change in land ownership. If the application is accepted, the governments register information about the change in ownership. Next, the governments can either withdraw the land for public conveyance or administratively allocate it to the village, and villages can choose to either keep or transfer the land to other parties. Regardless of the type of transactions, villages and the governments share the revenue; the ratio was close to 4:6 between landholders and governments at that time, subject to specific negotiation process.

(2)   TOR procedures

The redevelopment process involves various stakeholders: municipal government, district government, existing landholders (old factory owners, residents in old neighborhoods, village collective organizations, and villagers in old villages), and developers. Figure 6.1 reveals the TOR procedures for TOR application, examination, and implementation. Original land users are motivated to initiate renewal, and fiscally strong developers are selected as third parties to participate in the redevelopment process. All land subject to the TOR policy must go through a rigorous registration and application process so that actual property owners (individual or collective) can be identified (Ye, 2011).In the initial phase of the TOR process, the street community (*Jiedao* in Chinese) and town government examine old dilapidated neighborhoods, villages, and factories within its jurisdiction that do not comply with safety and environmental protection standards. Land parcels that qualify for the TOR program are identified by the district and

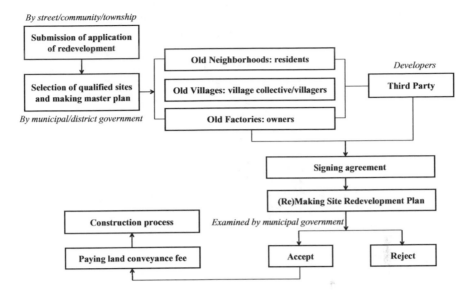

Figure 6.1 TOR procedure
Source: Tian and Yao (2018)

Guangzhou municipal governments, and a municipal-level renewal plan and several district-level renewal plans are made by the municipal and district governments, respectively.

The old neighborhood and old village land users are quite diverse. There may be hundreds of property owners in an old neighborhood, and it is quite difficult for them to achieve consensus on renewal. Similarly, there are a large number of villagers in an old village. Nevertheless, in most cases, the village collectives represent the villagers. Compared with those in the old neighborhoods and villages, the number of property owners in old factory areas is much less, and sometimes there is only a single owner. In order to save negotiation and time costs, the renewal of old factory areas is most popular among developers, followed by renewal of old villages. The most challenging renewal is that of the old neighborhood.

The policy for old factory renewal is different from that of old neighborhoods and villages. If factory use is changed for commercial, office, or a higher level industry, the factory owners can be granted self-redevelopment rights. The owners must then pay a land conveyance fee calculated on the benchmark premium differential between existing industrial use and new commercial or office use, and the upgrade of industry use is exempt of any land conveyance fees. If the industrial land is redeveloped for commodity housing, the redevelopment must go through the auction, bidding, or listing process, and the land conveyance fee is shared between the government and factory owner.

After listed by the TOR plan, current land user representatives (village collectives and neighborhood communities) must acquire permission for redevelopment from property owners. They then look for interested developers to conduct joint redevelopment, with the joint party responsible for calculating compensation and drafting the site redevelopment plan under the guidance of upper level plans. Government oversees these plans to ensure that improved land-use efficiency does not jeopardize livability in the city, the renewal is financially feasible, and the interests of existing property owners are protected (Lin, 2015).

When the redevelopment plan is examined and approved by the Urban Redevelopment Bureau (URB) and the joint party pays the land conveyance fee, construction begins. If the plan is rejected, the joint party has to redesign the site redevelopment plan. During the plan-making and approval process, the Guangzhou government, for example, introduces two rounds of stakeholder's participation in order to achieve a consensus. The first survey is conducted once the area to be redeveloped has been identified, and only after a certain percent positive response rate[1] has been achieved, can the process move ahead. The second round survey focuses on the compensation scheme and the detailed technical redevelopment plan. A minimum of two-thirds of all residents must approve the plan to start the implementation phase of the redevelopment (Ye, 2011).

In order to make the urban environment more livable and to protect public interests during the redevelopment process, the TOR policy requires a certain amount of land contributed for public use. Figure 6.2 presents land composition in the old neighborhood and village redevelopment. After the original construction is cleared, the site will be divided into three parts: Site1 is used for real estate development, and the land conveyance fee for this area is shared among the original landholders and government, and in many cities, this share is divided 60% (landholders) to 40% (government); Site2 is land for infrastructure and open space. The TOR policy requires the joint party should contribute at least 15% of the site land area for public use free of charge; Site3 is land for relocation of the original residents.

6.1.3 Power-relations, cost–benefits of stakeholders in the TOR

Table 6.1 presents the power-relation dynamics of interest groups between the traditional redevelopment approach and the TOR approach. As the most powerful agent in urban redevelopment, local governments initiated the institutional change and granted self-redevelopment rights to existing landholders. The framework of TOR seems like a win-win situation, and all stakeholders, including local governments, existing landholders, and developers, are supposed to benefit from the reform of the traditional redevelopment approach. In reality, implementing institutional change is not without costs. Table 6.2 lists the role, costs, and benefits of key stakeholders in the TOR process.

Site 1: Land for real estate development and land conveyance fee shared between government (40%) and existing land users (60%)
Site 2: Land for public facilities, road and open space, provided by developers
Site 3: Land for relocation, and existing land users receive compensation from developers

Figure 6.2 Land composition in the old neighborhood and village redevelopment
Source: Tian and Yao (2018)

Table 6.1 Comparison of traditional renewal and TOR approaches

Type	Traditional Urban Renewal	TOR
Initiating party	Local state	Existing land user
Role of the state	Dominant	Limited to examination and approval of redevelopment plan
LURs conveyance fee allocation	The state pays compensation for original land users and keeps the residual profit	LURs fee shared among existing landholders (60%) and local state (40%)
Means of LURs transfer	Bidding, auction, and listing by the state land banking center	Existing land user seeking third party for joint-development
Compensation for existing landholders	Displacement after cash compensation	In-kind compensation in the site or cash compensation depending on landholder preference

Source: edited according to Tian and Yao (2018)

(1)   Local government

The government provides local regulations to implement the TOR policy, makes macro-level renewal plans, and examines the detailed final redevelopment plan. Compared with the traditional redevelopment approach, local governments obtain less revenue from conveyance fees generated via LURs. However, they do not have to spend time and money on negotiating with and relocating existing landholders which eliminates high administrative costs.

Table 6.2 Analysis of cost and benefit of stakeholders in the TOR

Stakeholders	Role	Cost	Benefit
Local government	Policymaking; Macro-level plan making; Examination/approval of redevelopment plan	Profit concession of LURs fee Assuming risks in the land market	Obtaining 40% of LURs fees; Saving administrative cost; Industry upgrading;
Existing landholders	Initiating party	Concession of partial land for public use	Retaining 60% of the LURs fees
Third party/ developers	Joint development	Paying the financial cost for redevelopment; Bearing market risks	Acquiring LURs through negotiation; Profit sharing with existing landholders
Public	–	Pressure on infrastructure capacity induced by high-density redevelopment; Space gentrification	Land for public facilities; Improvement of environmental quality

Source: edited by the authors

Meanwhile, due to a relaxation in the land supply monopoly in the primary land market, local governments have had to assume the risk of oversupply in the real estate market induced by high-density redevelopment.

(2)   Existing landholders

Under the TOR process, 60% of the LURs conveyance fees go to the original land users, a much larger amount than they would keep under the traditional redevelopment approach. Current residents usually receive in-kind compensation at the site and do not have to be displaced to other areas, making relocation more acceptable. Besides in-kind compensation, residents can also choose monetary compensation with a lump-sum payment for their former house from the developer engaged in the joint-development. As part of the cost for self-redevelopment rights, original land users have to contribute at least 15% of land for public use such as roads and green spaces.

Formalization of rural collective landholdings strengthens the power of rural collectives over land. The TOR program represented a top-down recognition of legal rights for rural collectives to already converted rural construction land. Such an institutional change made rural construction land an asset of rural collectives and the potential land value could be fully capitalized through various market channels. Besides the gradual clarification of land rights between governments and rural collectives, there was also a process in which individual households (or villagers) were entitled more rights. As land redevelopment and land ownership change are both vital to rural collectives, cooperative members as the principals are given even more decision-making power in order to keep redevelopment of rural construction land flowing smoothly and to avoid social conflicts.

(3)   Developers

Under the TOR policy, developers can obtain development rights through negotiation. The TOR policy provides developers opportunities to acquire LURs at a lower land cost than through the traditional redevelopment approach where developers can only acquire LURs through auction, bidding, or listing. Developers under the TOR policy share redevelopment profits with existing landholders. Meanwhile, developers must bear all financial costs and assume risks, such as the possibility that the redevelopment plan may be rejected.

(4)   The public

In the TOR process, the public is like an "outsider" and is excluded from the decision-making process due to the lack of a public participation mechanism in China. On the one hand, the public benefits from improvements in environmental quality after implementation of the TOR; on the other hand, the public has to bear the pressure caused by high-density redevelopment, for example, traffic congestion and increased population, which in turn can bring about an insufficient amount of public facilities.

6.2   Constraints over TOR: multiple perspectives

6.2.1   Fragmentation along with three dimensions: increasing transaction costs

(1)   Transaction costs in land redevelopment

Transaction costs exist in the internal and external transactions of organizations, which are defined as search and information costs, bargaining and decision costs, and policing and enforcement costs (Furubotn & Richter, 2000). As for an organization, transaction costs are the costs incurred in transactions other than physical production costs (Cheung, 1987). High transaction costs pervasively inhibit exchange, production, and economic growth (Coase, 1960; North, 1991; Benham & Benham, 1997; Cheung, 1998). To a great extent, the degree of transaction costs is determined by the institutions which are created by human beings to reduce uncertainty (North, 1991, 2005). Efficient institutions which greatly reduce uncertainty can contribute to lower transaction costs and help make market transactions more efficient (Webster, 1998). However, as Buitelaar (2004) highlighted, not all existing institutions lead to and move towards lower transaction costs.

Whether lower transaction costs occur depends on the specific institutional arrangements and product markets.

Land is a special commodity in the marketplace because of its fixed location, heterogeneity, and extreme scarcity. The land market is characterized by fierce competition and extensive externalities, and thus, relatively higher transaction costs. Land development is a process in which stakeholders interact with respect to the transformation of bundles of rights and the value of land and buildings further shapes development outcomes (Healey, 1992). The land development process can be conceptualized as a series of transactions among stakeholders.

Generally speaking, land redevelopment in China involves at least five transaction stages: land-use planning, land assembly, land transfer, land serving, and property construction. Transaction cost in land redevelopment is greatly determined by how a series of institutional arrangements shape the transaction relationships and behaviors of involved parties in and across these stages. Specifically, as mentioned by Lai and Tang (2016) and Buitelaar (2004), the degree of transaction costs in land redevelopment is affected by four factors which include: 1) How land rights are delineated and the information on such delineation; 2) The number of parties involved in the transactions; 3) The degree of conflicts of interests among involved parties when an agreement needs to be reached; 4) Other factors which may affect the duration of land development.

It is reasonable to infer that transaction costs would be higher and land redevelopment would be inhibited more when land rights are not well defined, protected, and enforced, when more parties are involved, and when the efficient mechanisms of alleviating conflicts are absent, etc. Fragmentation as a typical characteristic of a peri-urban area is among the factors attributing to high transaction costs and impediment of land redevelopment.

(2)   Spatial fragmentation along with three dimensions in TOR

Fragmentation exists along three dimensions in peri-urban areas driven by rural industrialization: 1) the first is fragmentation of physical landscape. As a city is composed of a large number of villages which develop land in their area in a self-contained manner, the jigsaw of villages where different kinds of land uses are highly intertwined makes the entire city a fragmented landscape; 2) the second spatial fragmentation comes along with the dimension of landholdings or land rights. Although administrative villages are the basic autonomous administrative units, the economic organizations of rural collectives in the PRD region are mainly established at village groups (or natural villages) which are the basic unit of economic entity and compact social relationships. A land patch, despite its spatial continuity at the administrative village landscape level, would be held by many smaller actual landholders. Besides, the LURs are also fragmented in terms of small sizes of single land uses and the variety of leasing terms of different land plots; 3) the third dimension is caused by the preferential policies included in the TOR program. The land patch, despite its spatial continuity, the same use, and the same landholder, would be divided into several parcels of land according to whether the preferential policy was available or not.

(3)   High transaction costs due to fragmentation and their impediment to redevelopment

It is well known that high-quality and efficient land uses are based on a fairly large coordinated scale of land development and reasonable land plot sizes. A highly fragmented landscape will complicate negotiation in redevelopment process and incur at least two difficulties:

1) It would make plan-making and land assembly more difficult. In order to make a feasible and useful redevelopment plan, various and numerous landed interests should be fully considered. Apparently, complexity in making such a plan increases with the number of landed interests. Besides, fragmentation makes land assembly necessary. A higher degree of fragmentation will complicate negotiation more, as it increases the number of stakeholders and is easier for a hold-out problem to emerge.
2) The second difficulty lies in the complicated transactions by fragmentation. In the context of land fragmentation, transactions occur between numerous suppliers and numerous demanders. For the demanders, they must search among various landholders for expected one. They must negotiate with numerous potential landholders including rural collectives and sometimes even a large number of collective members to reach and execute an agreement. For the suppliers, in order to redevelop land of a certain area, repeated negotiation among the collectives and between the collectives and their members will occur to reach consensus about redevelopment intentions, land uses of redevelopment, expected profits, and ways of distribution. The suppliers also should pay tangible and intangible costs in order to find expected demanders and reach an agreement. As long as land rent differential is large enough, a rational agent will carry out redevelopment. However, one of the most important prerequisites is low transaction costs. Otherwise, it will become the major impediment of land redevelopment.

6.2.2 Intact welfare-oriented cooperatives: involution of collective economy in redevelopment

LSCs as welfare organizations, with the intrinsic problems as mentioned earlier, lead to the involution of their land management and the collective economy. First, both the cooperatives and members highly rely on revenue from land leasing. The maintenance of stable land revenue and thus steady dividends to shareholders becomes the top priority of the cooperative committee in determining ways of land management. Second, due to heavy dependence on land rents, the cooperative has to adopt conservative strategies in land management. It is apparently shown by their risk aversion, and the management tactics of low investment and low return. Any opportunity that would be captured by taking risk, no matter how low it is, may be dropped. Short-term returns and welfare are preferred over long-term economic productivity.

Such a situation existed not only before 2008, but also after that time when redevelopment began. The TOR program has provided various means of collective land redevelopment, such as spontaneous redevelopment by the collectives themselves, or redevelopment with landownership change. However, dependence on land revenue by both LSCs and their shareholders has also affected the renewal process. The LSCs usually require that redevelopment should keep their extant collective income unchanged during redevelopment, and the future collective income flow increased after redevelopment. This requirement determines whether an agreement of redevelopment can be reached, and in what ways that redevelopment can be implemented, and whether to initiate spontaneous redevelopment or to change landownership for redevelopment.

As there is a low ratio of collective revenue retained for future investment, the cooperatives are usually unable to afford large-size spontaneous redevelopment. Although cooperatives can cooperate with investors to carry out redevelopment, ubiquitous constraints over the capitalization of collective land (like financing from banks) imply the limited capital investment of investors. The low investment and small-size redevelopment together result in limited improvement to the collective economy which seems to be trapped without external forces.

Changing landownership to bring in powerful developers is also constrained in the context of the welfare-oriented management of LSCs. Landownership change will make the LSCs lose the land forever. In order to meet the requirement of maintaining a stable income flow, LSCs usually require a certain amount of properties which can be leased out for rents as they did previously. However, it is well known that management of the properties is usually more complex than raw land leasing, and needs more investment. Given the risk aversion of cooperatives and the non-market-oriented management, LSCs seem to hesitate on whether to carry out such a redevelopment. In our field investigation in some redeveloped villages, village cadres commonly say they cannot bear the heavy expenses on and risks of vacancy in property management. They prefer leasing all properties out with lower prices to ensure stable income to taking risks for potentially higher income. This slows down the speed of redevelopment of villages. Involution of the collective economy seems to be apparent even in redevelopment times.

6.2.3 Ambiguity of land development rights between urban governments and village collectives: strategic competition for land rents

(1)   Land development rights delineation: its importance for land-use efficiency

Land development rights are the rights to develop land and thus change the form and substance of land. As land development is usually characterized by extensive externalities, a clear definition of land development rights is vital for efficient use (Zhu & Hu, 2009). It is commonly accepted that land markets are always mediated by a "visible" hand in addition to an "invisible" hand" (Zhu, 2002). Planning control as a policy intervention is such a visible hand, which helps reassign land development rights to individual landowners by designating land uses and development intensity in advance. Protecting land users against the detrimental externalities, planning control is also considered as providing land markets with certainty and leads to maximizing social benefits (Fischel, 1985). However, as a state intervention to the market, planning control defines an individual's rights by simultaneously restricting his/her other rights. As a result, planning control may also attenuate individuals' rights and lead to governments' rent seeking (Furubotn & Pejovich, 1972). Effectiveness of planning control in delineation of land development rights is premised on the notion of the governments as an umpire in making and implementing the plan (Lai, 1997).

Land redevelopment refers to redistribution of land development rights and land value which involves two aspects. The first involves the relationship between public (social) revenue, which refers to the maintenance of public environment, and individual (economic) profit, which refers to capitalized land rents reflected by the market price of land (Smith, 1979). Through designating land uses and development intensity to specific land plots, planning control defines development rights of the land plot to its owner and thus profits to individuals. Public revenue is thus the differential between the intended maximized profits and the defined individual profits. The capitalized rental value of a land plot is largely determined by the equilibrium of demand and supply of land as a commodity in terms of its use and floor areas as designated by the plan (Zhu & Hu, 2009).

The second lies in the distribution of capitalized land rents among landowners in China, including both villages and urban states. In general, villages benefit from collectively owned land and urban governments receive revenue from the conveyance of state-owned land. However, redevelopment of collectively owned construction land often involves a change in landownership. In that case, both parties share the land rents. The government plays dual roles: that of the guardian of public revenue and that of the sharer of land rents. How it balances these roles depends on its political and economic interests which it expected to realize through redevelopment and interactions with villages under particular institutions.

Citizen participation in both the creation of a city plan and public monitoring of the execution of that plan has not yet been established in China (Ng & Xu, 2000). Planning control would be at the discretion of urban government and its role as judge may be attenuated. Land development rights defined by planning control would be ambiguously delineated. Land rents may be left in the public domain and up for grasp (Barzel, 1989). Similarly, if the principle of profit sharing between the two is unclear, fierce and disordered competition for land rents would also emerge. Land would be guided to suboptimal uses, instead of its best uses (Zhu & Hu, 2009).

(2) Ambiguous land development rights in TOR: its impact over redevelopment

Land development rights have not been clearly defined yet between villages and the government. First, after clarification of landownership and LURs, rural collectives have a priority of decision-making on whether to redevelop the land or not and whether to change landownership. Rural collectives are subject to the agreement of a certain ratio of collective members, which can be extremely high, up to 90% in redevelopment of old villages. That is to say, rural villages control the headstream of land supply in the redevelopment process.

Urban governments hold the authority of planning control which determines land uses and intensity, and thus land value in the market. Besides, as mentioned before, urban governments also hold the rights to specify ways of land transactions available to villages when involving landownership change, especially for the redevelopment of old villages. Associated with the ways are the ratios of revenue distributed between the governments and the villages. Through planning control, regulating transaction manners and the distribution ratios, urban governments actually control the "amount" of realized land rents and the part shared by rural villages.

Land development rights distribution between rural villages and the governments is unclear. Fierce competition and complex negotiation between them are expected. Controlling the headstream of land supply, rural villages would decide maintenance of collective landownership or not for redevelopment. The former choice implies that rural villages can receive land rents forever, but released land rent differential is very little. If rural villages redevelop land by themselves, the limited investment indicates a low increment of return. If external investors are attracted, increase in investment may lead to a raise of return. However, as collective land can neither be redeveloped to commodity property and nor be mortgaged for financing in banks, released land rent differential would also be limited.

Landownership change for redevelopment could bring about tremendous increase in revenue. Rural villages will attempt to fully maximize the value of the one-time opportunity and capture the land rent differential as much as they can. Rural villages would bargain with the governments for more profitable land uses such as commodity housing and high-end office buildings, higher development intensity, or an even larger distribution ration of land revenue. Otherwise, they may not agree with any redevelopment scheme.

Rural villages' strategic competition tactics hinder redevelopment progress in two ways. First, reaching agreement for redevelopment will take a long time due to bargaining, which results in high transaction costs. Second, rural villages' redevelopment intentions will tend to be polarized. They will prefer to either slow down redevelopment without landownership change to gain relatively lower but stable land rents or maximize land rent differential through redevelopment for commodity properties with landownership change. These strategies will slow down the speed of redevelopment.

6.3 TOR in Nanhai district: exploring the characteristics and reasons of spatial lock-in

Nanhai district is a typical peri-urban area. As an experiment for the redevelopment of low-efficient collective construction land, Nanhai district was chosen by the Guangdong provincial government to pilot the TOR program in mid-2007. As TOR program respects the *status quo* of urban forms and landed interests formed, it is necessary to analyze first the trajectories of rural industrialization and the sequent constraints over redevelopment, and then how redevelopment has progressed.

6.3.1 The constraints of rural urbanization over redevelopment in Nanhai

(1) Spatial fragmentation at multiple dimensions

Self-contained land development of a large number of LSCs and opportunistic land expropriation by urban governments lead to a fragmented landscape of the whole Nanhai. In order to explore the spatial patterns at the real scale of land development, it is necessary to go inside the administrative villages. Taking NH administrative village as an example, in the mid-2000s, 12 LSCs were established including 11 natural-village-based LSCs and one administrative-village-based LSCs. After expropriation of 85.7ha land by the governments before the mid-2000s, these 12 economic entities managed the remaining land. With industrial development, LSCs' land leasing has led to the rapid conversion of agricultural land to construction land, which is shown by the fast increase of industrial land ratio of total land from 21.1% in 1994 to 58.6% in 2012. Agricultural land, in 2012, was only 10.8% of the total, with the rest being residential land (30.6%). The highly interpenetration of residential land and industrial land is prevalent at the natural village scale, which are commonly considered to be spatially separated to reduce negative externalities (Figure 6.3).

Besides the physical landscape, fragmentation in Nanhai is also reflected in land rights. The first is landholdings of entities. As shown in Table 6.3

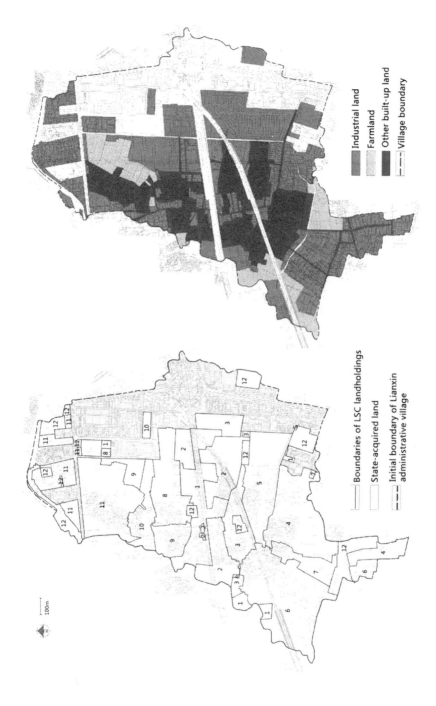

Legend (right map):
- Industrial land
- Farmland
- Other built-up land
- Village boundary

Legend (left map):
- Boundaries of LSC landholdings
- State-acquired land
- Initial boundary of Lianxin administrative village

N
100m

Figure 6.3 Land uses of NH administrative village and distribution of landholdings of entities, 2012

Source: drawn by the authors according to fieldwork

Table 6.3 Land uses of LSCs in NH administrative village, 2012

Codes for LSCs	Industrial Land		Farmland		Area of Village Residential Land (ha)	Total Land Area (ha)
	Area (ha)	Percentage in the Total Land Area (%)	Area (ha)	Percentage in the Total Land Area (%)		
1	6.8	52.3	1.1	8.5	5.1	13.0
2	14.1	72.3	0.0	0.0	5.4	19.5
3	8.0	65.6	0.0	0.0	4.2	12.2
4	13.6	51.7	5.0	19.0	7.7	26.3
5	11.6	61.7	0.0	0.0	7.2	18.8
6	15.4	53.7	6.1	21.3	7.2	28.7
7	4.8	68.6	0.0	0.0	2.2	7.0
8	8.6	64.2	0.0	0.0	4.8	13.4
9	6.5	63.7	0.0	0.0	3.7	10.2
10	4.3	52.4	0.6	7.3	3.3	8.2
11	12.4	44.1	8.9	31.7	6.8	28.1
12 (administrative village LSC)	11.9	72.1	0.0	0.0	4.6	16.5
Total	118.0	58.6	21.7	10.8	61.8	201.5

Source: Calculation according to fieldwork

and Figure 6.3, in NH village, a LSC holds land of 16.8ha on average, with the most being 28.7 ha, and the least being 7.0 ha. Industrial land is distributed across every LSC. An LSC holds industrial land of 9.8 ha on average. The left column in Figure 6.3 shows that almost every LSC's land disperses spatially to many intertwined parcels. The land held by the administrative village LSC is especially fragmented and dispersed in 11 parcels. The irregular boundary, distorted shapes, and interpenetration of land parcels of the entities lead to many mosaic parcels and non-continuous parcels. Also, the LSC, no matter how small its land area, has developed its land for industries and villagers' houses. Different types of land coexist in almost every LSC. As the basic unit of land management, these LSCs act as a "cell" with relatively complete functions. The cells coalesce to form the landscape of administrative villages, and the jigsaw of the cells of administrative villages then forms the entire landscape of the city. Spatial fragmentation and the large mix of conflicting land uses are not conducive to a living environment, especially when the population density is as high as 16,300 per square kilometer.

In addition to the fragmentation of landholdings, land rights are also fragmented in terms of plot size of single use and land leasing terms. In the entire territory of NH village, there were 575 industrial enterprises in 2008.

These enterprises consisted of labor-intensive industries such as furniture-making, textiles, and software. On average, each enterprise occupied a plot of 689 m². More than half of the enterprises occupied a plot with a size of 450 m². The extremely small sizes of the land plots used by low-end enterprises show the fragmentation of industrial LURs. Additionally, the leaseholds were different among these various enterprises. The shortest leaseholds were one to two years, the longest is 25 years, illustrating the fragmentation of the lease-holds. The fragmentation of LURs and leaseholds imply time-consuming and high costs for land assembly.

The third dimension of spatial fragmentation goes along the line with preferential and available redevelopment policy. As mentioned before, the TOR program aims to promote the marketization of collective construction land and transactions of land rights through clarification of landownership and other pertained bundles of rights. However, this is not meant to completely loosen the administration over redevelopment. That is to say, preferential policies associated with the TOR program have many prerequisites and are only available to those land plots which meet certain requirements.

In general, a TOR project needs to go through the following procedures step by step: 1) Land for redevelopment should be included in the provincial redevelopment database of Guangdong. Every half year, the database should be updated. Land entered into the database should be in accordance with land-use planning in effect and simultaneously recorded as construction land in the second land resource investigation in 2009; 2) Land should be included into the TOR plan which is made every five years by Nanhai district government based on the database. According to the plan, annual implementation plans are made by district and town government, respectively; 3) Projects included in the annual plan should be examined by town governments, and then reviewed by the district government to ensure the projects are in accordance with urban planning in effect, and other plans; 5) Only the approved projects can enjoy the preferential policies.

As a result, a land patch, despite its spatial continuity, the same use, and the same landholder, would be divided into several parcels of land according to whether the preferential policy is available or not. In order to avoid industrial hollowing in redevelopment, Nanhai government planned to keep and upgrade several village industrial parks. In 2015, 673 village industrial parks were delimited, and the total land area reached 9,495.27ha. These parks were spatially dispersed, with an average size of 14.11ha, with a minimum of 0.18ha, and a maximum size of 188.35 ha. The average size is smaller than that of industrial land patches with the entirety of Nanhai as a landscape unit (18ha). As for total parkland, the land that agrees with the land-use plan of Nanhai's whole territory accounts for 91.35% (Table 6.4).

However, the land area included in the provincial database was greatly reduced relative to the land that confirms to the land-use plan, with the ratio to the total land area of village industrial parks at only 59.35%. The average land area that a village industrial park has been included into the

Table 6.4 Statistics of land in village industrial parks that meets requirements, 2015

Land Classification by Available Policies	Area (ha)	Proportion in the Total Land of Delimited Village Industrial Parks (%)	Mean Value of Land in Village Industrial Parks (ha)
Land of delimited village industrial parks	9,495.27	100.00	13.86
Land in accordance with land-use plan	8,673.72	91.35	12.66
Land included in TOR database	5,635.28	59.35	8.23
Land of recognized TOR projects	3,774.41	39.75	5.51
Land with legal land uses rights certificate	3,479.44	36.64	5.08

Source: Nanhai Land Resource Bureau

database is only 8.23 ha, nearly half of the average size of the village industrial park. The land area of recognized TOR projects further decreased to 3,774.41ha, which only accounted for 39.75% of the total area. The average of recognized redevelopment lands across village industrial parks was reduced to 5.51ha. Both the ratio and area of land that had certificates of LURs decreased further in contrast to the land of recognized projects. As only recognized TOR projects with certificates of land-use rights can enjoy preferential redevelopment policy, a land patch would typically be subdivided into pieces spatially dispersed.

In sum, spatial fragmentation has historical causes, with the latter dimension of fragmentation dependent on the former. Government efforts to solve the former two dimensions of fragmentation lead to the spatial fragmentation along with available policy. The overlapping of three dimensions of spatial fragmentation makes integrated land redevelopment more difficult. Fragmentation essentially complicates the many-to-many land transactions, further leading to high transaction costs.

(2)   Ambiguous land rights inside cooperatives and between them and the governments

Although the TOR program provides channels to clarify the ownership of collective land and some associated bundles of rights, ambiguity of land rights also exists inside the rural collectives and between the LSCs and the urban governments.

First, as mentioned before, the open-membership cooperatives in Nanhai suffer from the intrinsic cooperative problems which result from the unclearly defined residual claimant right of members, and the absence of a free market of shares transaction. The number of shareholders rapidly increased from 583,100 in 2000 to 756,700 in 2008 (NHBDPS, 1998–2013). Members as principals tended to require short-term dividends allocation instead of long-term investment and the LSC committees as agents tended to be quite conservative. They usually require that governments keep their extant collective income flow unchanged and then compete for land rent differential with the governments. Collectives were disincentivized to invest to redevelop land by themselves, especially when the expenditure would affect their annual dividend allocation. In spite of self-redevelopment, the size of projects would be small. This was apparently shown by the smallest average size of projects redeveloped by LSCs (31.74 mu) in contrast to those invested by developers (59.03 mu), governments (45.08 mu), or both the developers and the governments together (642.84).

Second, land development rights are not clearly defined in Nanhai's TOR practices. While LSCs actually hold the land, the government holds the rights to determine the uses and intensity of land. The former bargains with the latter by means of *de facto* landholdings to maximize their profits, and the latter has its own interests and responsibility to maintain public profits. Initiation of redevelopment is built upon their cooperation, but strategic competition between them is inevitable, especially when involving landownership change. To change landownership, the LSCs will tend to maximize the profit in a lump-sum deal. Commodity property is preferred above all other uses. It is clearly shown that the area of redevelopment projects of commodity property accounts for 72.48% of the total redeveloped land area in Nanhai. However, appropriate locations for commodity property redevelopment are limited and the Foshan municipality government instituted controls on the overheated redevelopment of commodity property in 2012 in order to stabilize the real estate market. This explains, from the supply side, why Nanhai's redevelopment slows after that time.

To summarize, the clarification of the land rights associated with the TOR program is not thorough. Shareholder's claimant rights over collective revenue inside rural collectives and development rights between the collective and governments remain unsolved. Marketization of collective land is impeded because of the conservative behaviors and high expectations of profit of rural collectives.

6.3.2 *TOR in Nanhai: three phases and general characteristics*

(1)   TOR in three phases between 2007 and 2018

Generally speaking, TOR in Nanhai has experienced three phases according to the rhythm of redevelopment and general institutional circumstances: 1) the first phase existed from mid-2007 to mid-2009. During the time, TOR was experimented first in some cities, like Nanhai, and there was not any provincial policy guide. In essence, that was a phrase about the local practice of the TOR that reflected China's pragmatic reform philosophy, literally, "crossing the river by feeling the stones." Without clear guidance from the top, institutional uncertainty was quite apparent. In such an environment, the speed of redevelopment was relatively slow (Figure 6.4).

2) The second phase between 2009 and 2012 was the first experimental contract period between Guangdong province and the central state. The contract *per se* and the "Opinions" issued by Guangdong provincial government provide a relatively certain (or stable) institutional environment which incentivizes governments at lower levels and the landholders, and speeds up the renewal process. As shown in Figures 6.5 and 6.6, there was, by and large, a trend of increasing area for new TOR projects. As seen during this phase, institutions also changed very frequently. In Nanhai district, more than 20 regulations have been issued based on the "Opinions." And most of the regulations changed every year. Moreover, the "Opinions" highlights some provisions which were only valid until 2012. One such provision regulated formalization of illegal land uses for parcels included in the TOR program. Since formalizing illegal land uses was a prerequisite of renewal, the 2012 deadline led to a wave of hasty applications of land-use formalization and renewal projects for fear of losing out on preferential policies. In sum, the top-down recognition of the TOR and the future uncertainty of preferential policies together led to an annual increase of the area of new TOR projects.

3) The third phase was the period from mid-2013 to present. After the first contract expired at the end of 2012, there was a short period of institutional uncertainty, as the effects were not announced by the central state until the middle of 2013. As a result, both the area and numbers of new renewal projects witnessed a severe drop in 2013 in contrast to 2012 (Figure 6.4). In June 2013, the Ministry of Land and Resource approved that the contract

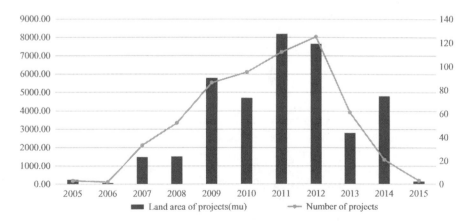

Figure 6.4 Land area and number of renewal projects that had begun construction
Source: Nanhai Land Resource Bureau

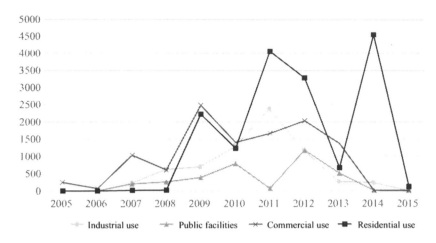

Figure 6.5 Changes over years of the area of various land uses after renewal, 2005–2015

Source: Nanhai Land Resource Bureau

period of TOR with Guangdong province was extended to run between 2013 and 2020. This was a very important event which deeply affected the progress of the TOR program, because the agreement with the Ministry of Land and Resource explicitly implied the TOR would be a long-term program. A relatively stable institutional environment had now been established, and this provided local governments and landholders more time to evaluate the practical effects of previous policies and thoroughly explore ways of sustainable urban renewal. As a consequence, local governments' expectations tended to now be rational rather than hasty renewal. Shown in Figure 6.4 was the decrease of the number and area of renewal projects in contrast to the first contract period.

(2) TOR program: steady progress in face of unprecedented challenges

Generally speaking, Nanhai's TOR made some progress. However, when compared with the large amount of planned re-developable land, the renewal only took a small step forward. From 2007 to the end of 2015, the number of already implemented TOR projects (including those that had already been completed or had begun construction) was 598. The total redeveloped land area reached 37,393.27 mu (Table 6.5). Meanwhile, the total land area included in the provincial database was 296,864.36 mu,[2] and the area of land parcels recognized by the district government for renewal was 167,900 mu. The total redeveloped land area accounted for 12.6% of the total land area in the database and 22.28% of the recognized renewal

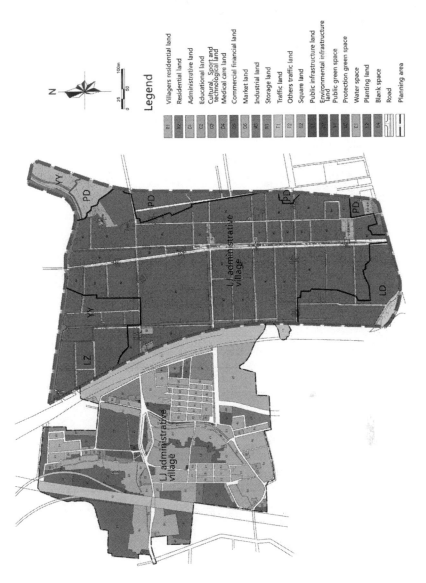

Figure 6.6 Land uses and landholdings in LJA, 2007
Source: edited according to Yuan et al. (2015)

land area, respectively. These percentages show the renewal plan for the next 5–10 years has, to some extent, begun steadily. As Table 6.5 shows, the land area of implemented renewal of collective-owned and state-owned land accounted for 11.94%and 14.52%, respectively, of the total land area in database, indicating that the TOR program still faces many challenges.

Table 6.5 Progress of TOR program: land area and ratios up to 2015

Types of Landownership and Uses		Total Land Area in 2007 (mu)	Land Area in TOR Land Database till 2015 (mu)	Implemented TOR Projects up to 2015	
				Land Area (mu)	Ratio to the Land Area in Database (%)
Collectively owned land		1,118,168.66	221,012.24	26,380.04	11.94
In which:	Old factories	361,185.20	136,805.70	16,418.75	12.00
	Housing sites	165,619.40	64,692.64	2,594.28	4.01
	Farmland	560,608.58	16,645.27	3,375.88	20.28
	Other construction land	30,755.48	2,868.63	3,991.13	139.13
State-owned land		480,423.63	75,852.12	11,013.23	14.52
In which:	Old factories	129,351.91	40,916.94	7,350.71	17.96
	Old town areas	67,930.02	17,590.59	1,595.12	9.07
	Other land	283,141.70	17,344.59	2,067.40	11.92
Total		1,598,592.29	296,864.36	37,393.27	**12.60**

Sources: Nanhai Land Resource Bureau

(3)   Demand-side: property-led redevelopment and industry-oriented renewal

Land will be redeveloped only when the land rent differential is large enough. Accordingly, commodity housing and commercial properties have stronger ability to bid for rent, in contrast to industrial uses and public facilities. As shown in Figure 6.5, residential and commercial use after renewal accounts for 72.48% of all redeveloped land, the industrial land area after renewal only accounts for 18.36% of the total, and the land area for public facilities only takes the share of 9.16%.

Such market-driven and property-led redevelopment, however, experienced a downward trend after 2012 before which there had been a fever of residential and commercial redevelopment. The reasons might be: 1) there was an overall downward trend of renewal after the time; 2) although residential and commercial land redevelopment could release huge land rent differentials, these land uses tended to rely on good locations which were finite; 3) in fear of an overheated property-led redevelopment and the resultant hollowing out of industries, in particular, manufacturing, the Foshan municipality government issued regulations to encourage industrial upgrading and cool the fever of residential redevelopment. This is evidenced by the change in the distribution proportion of land revenue for property-led redevelopment between the governments and landholders from 6:4 to 7:3 at the end of 2011. Meanwhile, the municipal government required every district to have at least 60% of old factories redeveloped to industrial premises and for them to surrender part of the land revenue to landholders and to reward enterprises. Despite the encouragement from governments, industrial redevelopment was not preferred by the market.

(4) Supply-side: collective land and old factories as the main redevelopment objects

In Nanhai district, rural industrialization has led to a huge amount of inefficient collective land which became the main source of TOR projects. As Table 6.5 shows, of the total redeveloped land area between 2007 and 2015, collective land accounted for 70.55%, with the land area of 26,380.04 mu, and the remaining 29.45% was state-owned land. The collective land was nearly 2.5 times the state-owned land. Collective industrial land accounted for 62.24% of the total redeveloped collective land. Data in 2015 revealed that the number of old factories, old villages, and old town area renewals were 349, 107, and 142, respectively. The land area of the three types of renewal accounted for 65.88%, 12.25%, and 21.87%, respectively. Therefore, old factories were the main source of TOR projects.

The comparison of land-use change before and after renewal presents that the amount of land area converting from industrial to residential and commercial uses was 25,647.95mu, and the area of industrial land without change in uses was 6,853.67mu. The area of residential and commercial land without change of uses was 4,621.54mu. The remaining land was 270.11mu (Table 6.5). It is apparent that change from industrial to residential or commercial uses was the main type of TOR, which brings about huge land rent differentials.

(5) Driving forces underlying renewal: the vital roles of capital-holders

To capitalize land rent differential through market channels is one of the intentions of governments in carrying out the TOR program. As landed interests have formed, including individual households, rural cooperatives, land tenants, and even local governments, successful redevelopment projects tend to be those that make all the interests content with the distribution of land revenue which depends heavily on whether capital holders are willing to invest.

It is evident in Nanhai's TOR program that participation of capital holders (developers) plays a key role. In general, there are six types of funding sources for redevelopment. They consist of rural villages as single investors, governments as single investors, developers as single investors, joint investment by rural villages and developers, joint investment by governments and developers, and other forms of investment. Data show that the number and land area of renewal projects invested by developers alone are 353 and 20,859.07mu, accounting for 59.3% and 55.78%, respectively, of all the implemented projects (Table 6.6). The number and land area of renewal projects invested jointly by

Table 6.6 Information of projects funded by different investors

Items \ Investors	Villages	Government	Developer	Villages and Developers	Government and Developers	Others	Total
Land area (mu)	3,999.42	991.75	20,859.07	2,462.82	7,071.27	2,008.94	37,393.27
Ratio of land area in the total(%)	10.70	2.65	55.78	6.59	18.91	5.37	100.00
Number of projects	126.00	22.00	353.00	37.00	11.00	49.00	598.00
Ratio of numbers in total(%)	21.07	3.68	59.03	6.19	1.84	8.19	100.00
Land area of a project(mu)	31.74	45.08	59.09	66.56	642.84	41.00	–

Source: Nanhai Land Resource Bureau

developers and governments, and by the developers and rural villagers are 48 and 9534.09mu, accounting for 8.03% and 25.50%, respectively, of all the implemented projects. If considering all the projects in which developers are involved as investors, the ratio of the number and land area are 75.25% and 86.65%, respectively. There are 126 TOR projects invested in by rural villages alone, which accounts for 10.70% of the total redeveloped land area. There are only 22 renewal projects solely invested in by governments, and the land area accounts for 2.65% of the total redeveloped land.

In sum, Nanhai has made great progress in redevelopment of the collective land, but still faces great challenges under a series of constraints among which fragmentation and LSCs' strategic competition matters much. The following two sections will explore in detail how governments act to conquer fragmentation problems and, compete and cooperate with LSCs and developers through two case studies. One case study is a project in an industrial district involving land integration among administrative villages. The second case study is a redevelopment project with an administrative village as a whole and coordination mainly happening among natural villages.

6.3.3 Integrated redevelopment across administrative villages: LJ industrial area (LJA)

(1)   Background: the first integrated redevelopment project in Nanhai district

LJA is an area with a territory of 122.49 ha in Nanhai's DL Town, and is located in the eastern part of the district between the urban cores of Guangzhou and Foshan. The whole area consists of five administrative villages

(19 natural villages),[3] with LJ, LD, PD, YY, and LZ having a land area of 83.77ha, 11.15ha, 9.57ha, 10.17ha, and 7.83 ha, respectively (Figure 6.6). Since the early 1980s, garbage collection, processing, and recycling activities had gradually concentrated in the area, and the LSCs leased their land out to accommodate a large number of garbage-related enterprises and workers.

At the end of 2006, Sky News reported that LJA had become a recycling center for British plastic waste and, people as young as 14 made a meager living by reprocessing mountains of toxic rubbish (http://news.sky.com/home/article/ 13553680, accessed 29 July 2012). After news reports surfaced on the serious environmental pollution, Nanhai local governments immediately clamped down on the waste recycling business and, consequently, decreased the rental income flow of the LSCs. To maintain the LSCs' collective income and promote development of LJA, the DL government began their redevelopment program at the end of 2006.

If the report by foreign media helped ignite redevelopment, the change in external location conditions made it a necessity. First, before 2005, LJA was situated at the juncture of three smaller towns which merged to form the DL town in 2005. After the merge, LJA became the geometric center of the new town (DL Town). Second, with the location advantages improved greatly, LJA was then designated as a commerce and trade center and the Non-ferrous Metal Headquarters base in the Nanhai's master plan. Third, the advantageous location highly increases land rent differentials which is shown by the former monthly industrial land rents (3.5 Yuan/m^2) and the potential property rental of business offices (25–35 Yuan/m^2). The increased land rent differential promoted redevelopment. The redevelopment of LJA is a pilot project to lead the TOR program in Nanhai district, before the issue of Nanhai's official guideline of TOR.

(2)   Redevelopment in three stages

Although the land rent differential has facilitated the renewal process, the TOR has encountered several barriers. The first is how to draw out an attractive and feasible redevelopment scheme to gain agreement from involved LSCs who are the landholders. The second is the fragmented land ownership and irregular land boundaries of different landowners. Additionally, as location conditions and land rents are different among LSCs, how to make a balance among them with respect to sharing land profits after redevelopment or possible land adjustments is also in difficulty. In order to overcome these difficulties, an exploratory, tentative, and pragmatic redevelopment program has been gradually carried out in three stages.

The first stage created a redevelopment scheme that could be agreed to by all LSCs with respect to compensation to the landed interests and sharing of future land rent differential. It nearly took one and half years, but a contract between the LSCs and government was reached by the end of 2008. According to the contract, the government compensated enterprises for their moving, and paid LSCs for their loss of land rents in 2007 and 2008. The yearly payment to the LSCs amounted to 16 million Yuan. As to profit-sharing in the redevelopment process, the government and the LSCs reached an agreement: the government would rent all the land in LJA on Jan 1st 2009 for 40 years at a monthly rental of 3.5 Yuan/m². The monthly rent would increase by 10% every three years. For planned roads, public facilities, and space, the government would rent land from LSCs and invest in their construction.

After the contract period ends, if the LSCs want to maintain collective land-ownership, the government can sublease land to investors for redevelopment with a leasehold of 40 years. After the lease expires, properties built by investors should be handed over to the LSCs, and the government would stop the monthly rental payment. As only state-owned land can be used for tradable commercial office building and housing, LSCs can apply to the government for landownership change for such uses after reaching an agreement for profit sharing between the government and the LSC. After a plot of land is conveyed to developers, the government would stop rental payment for the plot.

Compensation and profit-sharing arrangements removed the LSCs' discontent, and eliminated their worries of future uncertainty. Landownership changes determined by the LSCs entitled them more power to negotiate with the government for sharing land rent differential and bargain for wider distribution. These positive expectations incentivized the cooperation of LSCs in the redevelopment process, making a good start for continued redevelopment negotiations.

After reaching agreement on redevelopment based on a commonly accepted overall profit-sharing principle, the second stage was to negotiate among the LSCs for land consolidation which is necessary for the following three reasons: 1) LSC lands are highly intermingled with irregular shapes, which is not conducive to the overall development of LJA; 2) according to the master plan of LJA redevelopment (Figure 6.7), the government needs to expropriate a land plot with an area of about 14ha at the south-west corner of LJA. This land would be used for high-end business office buildings and luxury hotels to develop a headquarters economy. This would help improve the city image of LJA, attract talent, and push forward the redevelopment of entire area. The strategy adopted to attract investors was to lower land prices, which further needs to pay lower compensation for land expropriation. Early government compensation and investment won the LSCs' agreement of land expropriation at a lower price. However, expropriated land should be apportioned among the LSCs; 3) as different uses have been designated for each of the land plots, it means different ways of land transaction and thus different land rent differential. Equality and justice among LSCs requires the reallocation of LSCs' land according to their redevelopment intentions and proposed land uses.

6.3.3.1 Land use and land adjustment of LJA

At the end of 2006, the entire LJA was plotted out according to the planned roads (Figure 6.7). Land ownership boundaries were also redrawn after

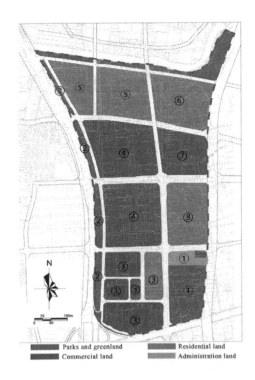

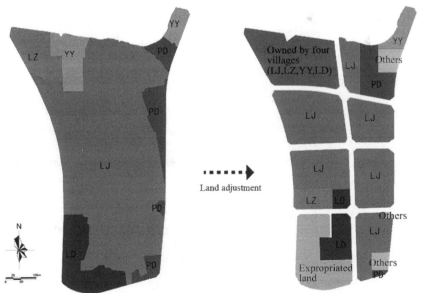

Figure 6.7 Master plan of land use and land adjustment based on the plan of LJA
Source: edited according to Yuan et al.(2015)

considering a series of elements such as, LSCs' land-use preferences, different location conditions, and original land areas included in LJA. After land adjustments, each LSC held several land plots with clear boundaries and regular shapes. The certificates of ownership and use rights of land plots were also issued to the LSCs. Through adjustment, every plot of land had clear landowners and the involved landowners reached consensus with respect to land uses, potential land transaction manners, and possible profit-sharing principles. Preparations were made for developers' participation in the redevelopment process

If the first two stages made all necessary preparations for redevelopment, the third stage broke institutional constraints, and explored various modes of land transactions and redevelopment based on the principle of "one plot one policy." Combining the elements of land use, project characteristics, and the intentions of the government and LSCs, LJA adopted five new land redevelopment modes in terms of land revenue distribution and manner of land transactions (Figure 6.7, upper).

1) Plots 1 and 2: These two plots were used for public use like parks, green space, and roads which are fundamental for the improvement of urban spatial quality. However, because of its nonprofit characteristics, owners of collectively owned land are usually not willing to supply land at no cost. To expropriate land at market price was beyond the government's fiscal capacity. Under this backdrop, the government rented land from LSCs and invested about 100 million Yuan in advance to construct facilities.

2) Plot 3: This plot was used for the Non-ferrous Metal Headquarters base. As mentioned before, this plot was expropriated by the government and then conveyed to investors. The price of land expropriation and LURs fee is low as due to the intention of the governments to reshape city image and attract large and high-quality enterprises through flagship projects.

3) Plot 4 was used for wholesale and retail service. The plot was leased by the government and then subleased to investors for a 40-year term, and collective land ownership remained unchanged. In this context, investors tended to be a property manager, instead of a real estate developer. The government played a vital role as a "transaction broker" between LSCs and investors.

4) Plots 5–8 were for residential and commercial uses with the land ownership changed from collective to state-owned. The procedures for converting land to be state-owned were completely different from traditional land expropriation under which the state acquires land from the collectives after paying one lump-sum compensation. First, the LSCs applied to the government for a landownership change after receiving approval of two-thirds of cooperative members. The government then registered the landownership change and allocated

the land at no charge to the LSCs. The LSCs then openly transferred the state-owned land and shared land revenue with the government. These procedures were intended to grant the decision-making power of land transactions to landowners in order to incentivize rural collectives to supply land, but the profit-sharing models varied case by case.

Over the past ten years, redevelopment of the entire LJA has been nearly finished. Of the total land area, 42.91% of land (52.55 ha) changed land-ownership and the remaining 57.08% (69.92 ha) has remained collectively owned.

(3)   Negotiation-based coalition led by local governments

The above redevelopment process shows an unprecedented change of trad-itional governance over land development. Before the TOR program existed, governance was either coercive land expropriation by urban governments or absence of government supervision over rural collectives' spontaneous land development. Previously, there had been resistance to redevelopment or competition between rural collectives and urban governments with respect to profit distribution in land development. However, with the TOR pro-gram, rural collectives hold land ownership and use rights which bring them rental income, but the government holds land development rights which determines future profit. Cooperation between them is the only way to capture future profit. This gives birth to a new governance of coalition between the two parties. In LJA, the coalition starts from a contract which establishes a new commonly accepted profit-sharing mechanism. The con-tractual relationship between the government and the LSCs breeds mutual trust with respect to future redevelopment. Both the government and LSCs agree to and actively promote redevelopment to capture the expected huge profits. It is evident that formation of this profit-centered coalition is based on the leading role of urban government. It recognizes and compensates the LSCs' vested profits and agrees to share future profits with them.

However, the profit-centered coalition is negotiation-based, shown by the last two stages of redevelopment, which exhibits both cooperation guided by the government and competition for land rents mainly sourced from LSCs. Cooperation is shown by the agreement and efforts made among the government, the LSCs and even the investors, to reduce transaction costs to promote land transactions. In the cooperation, the government's central role as a "transaction broker" between supply and demand sides should be high-lighted. On the supply side, local government persuades LSCs to supply land through a new institutional arrangement outlined by the contract. It then coordinates with the LSCs to adjust fragmented land to make the plot regular. On the demand side, it rebuilds confidence in the market by show-ing a magnificent blueprint of plan, making preferential policies to lower the land cost and substantial investment in public facilities and spaces. In the face of distrust between the LSCs and investors, local government guarantees trust to both, as a broker, and promotes the transaction.

Competition is also a form of negotiation-based coalition in LJA redevelopment, demonstrated mainly by LSCs' strategic competition for land rent differentials and the corresponding compromise of government. Although the basic principle of profit-sharing is clear, negotiation for land transactions and profit-sharing is plot-based, causing an institutional uncertainty. Despite this uncertainty, the government made large investments in order to initiate the redevelopment. However, the government was finally held out by the LSCs. First, the LSCs fully expressed their requirement that redevelopment should not affect their vested income flow and ensure the sharing of land rent differential(s). Second, with the success of early-stage projects, LSCs gradually required more profit-sharing. At the beginning of the process, LSCs accepted allocation of 70% of land conveyance fees. However, for the last three plots, numbers 6–8, they required 35% of the total properties in addition to 70% of the land conveyance fees.

For the government, its first objective is to ensure the successful implementation of the plan, which can improve city image, upgrade industries, and provide a firm basis for future tax flow. Instead of maximizing land revenue as the government usually did before, it compromised in land revenue and made profit concession to promote redevelopment. A rough estimation shows that LSCs together received 3.399 billion Yuan from the redevelopment, which was far beyond the discounted value of their former annual land rents (1.05 billion Yuan). The government received 1.613 billion Yuan, and spent 1.204 billion Yuan. The net income ratio between the government and LSCs was 85:15. If the required 35% of properties were included, the LSCs would be better off than they were previously. The government's concession of profits to the LSCs facilitated the renewal process.

While land rent appreciation was largely due to the government's investments, land value capture should be shared among the public. In reality, however, the LSCs and developers were better off at the expense of public expenditures, and the public were excluded from the renewal process.

6.3.4 Integrated redevelopment of an administrative villages: XB village

(1) Background of redevelopment

XB is an administrative village in Nanhai district which epitomizes the rapid urbanization driven by both urban state and village collectives. In the early 1980s, XB village covered a territory of 9,750 mu. The Nanhai government acquired some farmland in XB village before 1994, and collectively owned land was reduced to 3,765 mu (251ha); thereafter, it had remained essentially unchanged. In the mid-1990s, five affiliated natural villages (groups) and the administrative village committee established an LSC, respectively. Up to 2009, six entities had been set up to manage their land, with areas ranging from 21.2ha to 67.4ha (Figure 6.8 and Table 6.7). Every LSC as an autonomous and independent unit carried out self-contained land development for economic activities (mainly manufacturing) and/or housing for villagers. By 2009, XB village had attracted 382 small factories, including hardware and plastics industries, among others. It also accommodated 6,663 local villagers and approximately 12,000 migrant workers. Driven by rural industrialization and population increase, XB witnessed rapid farmland conversion. From 2000 to 2009, the ratio of construction land in the total increased from 39.8% to 80.9%, and that of industrial land rose from 18.0% to 49.3% and residential land increased from 21.8% to 31.6%. Factories, housing, and farmland were intensively mixed in every LSC, illustrating a deteriorating living environment and inefficient land use.

Location change with the change of city development strategy promoted the village redevelopment. In the course of its spontaneous non-agricultural development, XB saw its location changing from a remote rural area to an urban center, surrounded by major urban roads. Two metro stations, which are located on the west side of Nanhai, connected the village to the urban centers of Guangzhou and Foshan. Moreover, the district government designated XB as a future financial and business center. As a result, the land rent differential quickly increased. In 2010, the conveyance price per square meter of state-owned land

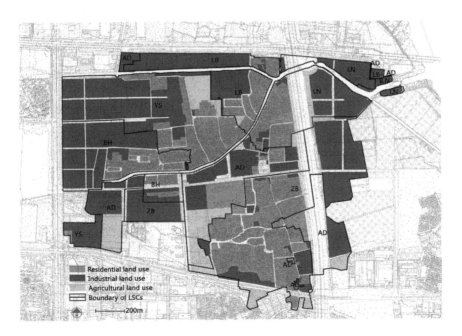

Figure 6.8 Land uses of XB village, 2009
Source: Nanhai Planning Bureau

Table 6.7 Statistics of built-up areas in LSCs, 2009

LSC	Residential Land Uses		Economic Development Land Uses		
	Land Area (ha)	As % of the Total LSC Area	Land Area (ha)	As % of the Total LSC Area	No. of Industrial Enterprises
BH	8.2	26.0	16.5	52.4	50
YS	15.0	32.2	22.1	47.4	77
LB	7.0	33.0	10.9	51.4	48
LN	8.0	26.1	12.4	40.5	38
ZB	28.8	42.7	20.0	29.7	116
Administrative village committee	3.1	6.0	27.3	53.2	48
Total	70.1	27.9	109.2	43.5	382

Source: Nanhai Planning Bureau

nearby had already reached 18,000 Yuan, whereas the discounted present value of yearly rents per square meter of industrial land was only 2,262 Yuan, and the large potential land value appreciation motivated redevelopment.

However, redevelopment faced the great challenge of fragmentation. In addition to the physical and landholding fragmentation, applicable redevelopment policies which vary from land use type to type also exaggerate the spatial fragmentation. For instance, industrial land redevelopment needed the agreement of two-thirds of all members. If a decision involved landownership change, the ratio of agreement of members needed to reach 90%. For residential land redevelopment, the threshold was even higher and necessitated a 98% agreement rate. In 2009, XB village was selected by the Nanhai city government as a pilot project of experimenting with integrated redevelopment of the entire village necessary for improving environmental quality.

(2)   Renewal of three phases in XB village

XB village redevelopment began in 2010 and can be divided into three stages. The first stage was to sign a "renewal implementation scheme" contract between the renewal office affiliated to the district government and all LSCs. The contract briefly regulates the entire period of renewal, roles played by the government and the LSCs, the renewal process, profit-sharing program, etc.:

1)   The contract period was set to be seven years, from 2011 to 2018, during which village redevelopment was expected to be finished.
2)   The government was to assume responsibility of coordinating overall redevelopment, including plan-making and its implementation, controlling

and adjusting the renewal progress, addressing emerging conflicts, and promoting land transactions. The LSCs determined whether to supply land for renewal and whether to change landownership or not.

3) Integrated redevelopment meant the government would negotiate with all LSCs at the same time to determine the renewal progress of the entire village. The plan called for old factories to be redeveloped first to obtain revenue to distribute to stakeholders and then to support subsequent renewal of village settlements, which were considered difficult to redevelop because of the large number of households involved.

4) The government invested in advance to balance the landed interests, make site preparation, and promote land transaction and renewal plot by plot. The government bore early costs such as compensating and removing the original tenants, site-cleaning and land-leveling, construction of infrastructure, and compensation to LSCs. The government committed to pay the LSCs for their lost annual land rental from 2011, maintaining a stable, unchanged flow of collective income during the renewal period. The initial rental was equal to that received by the LSCs from land tenants in 2010, growing at a rate of 8% each year. Once a plot was sold, the corresponding rental payment was terminated in the following year. For plots that remained undeveloped during the contract period, payment was renegotiated. If land ownership was changed, the government withdrew, and publicly conveyed the land, the revenue would be equally shared between the two parties. If land was administratively allocated to the LSC for transfer, the government received 40% of the transaction price.

After signing the contract, the second stage of renewal began in 2011. This stage includes redeveloping part of the old factories and redistributing land revenue sourcing from it. In early 2011, the district government raised up a start-up fund of one billion Yuan to compensate previous land tenants, pay yearly rents to all natural-village-based LSCs, demolish old factories, and build infrastructure. In order to address the fragmentation problem, the district government asked for agreement of the natural-village-based LSCs whose land plots were next to each other to change landownership for public conveyance, and then attempted to pool the land plots belonging to different LSCs together to attract large developers (plots 6 and 7 in Figure 6.9).

At this stage, 725.81mu of land in the XB village was conveyed, resulting in a total revenue of 8.6 billion Yuan, 40% of which was distributed to local governments, 10% went to the central state, 20% distributed to LSC members, and the final 30% was captured by the LSC committee. In general, local government played leading roles through investing in infrastructure, addressing the land fragmentation problem, and attracting large developers in the land capitalization in XB village.

Moreover, most of the designated land uses was for residential, office, and retail commerce use which maximized land rent differentials, and the higher floor-area-ratio ranged from 3.5 for residential uses to 5.8 for office buildings.

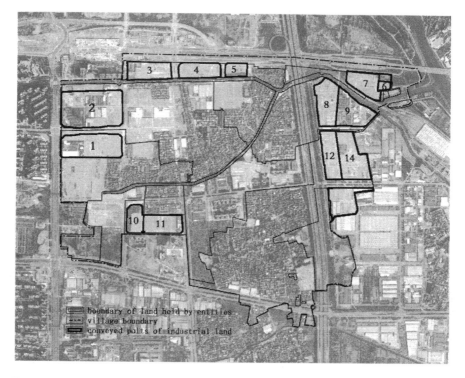

Figure 6.9 Conveyed land plots during the first-stage redevelopment
Source: drawn by the authors

The land income was mainly generated from the land rent differential due to the conversion of a large proportion of the old factory into commodity property.

The LSCs or the members preferred land conveyance in order to receive as much revenue distribution as possible. For BH and YS, both only conveyed a small proportion of their territory and they still had a certain proportion of leveled land for redevelopment. However, for LN, ZB, and LB, in order to allocate their members more revenue, most of their territory was conveyed. Nearly no leveled land was left in the three LSCs, making continuous renewal difficult.

In 2014, the government attempted to promote the third stage of renewal which was to redevelop the remaining land of the LSCs. However, the government finally found that the negotiation process became more complicated and faced harsh challenges. First, the government and the LSCs worked together to rebuild a certain amount of collective properties for the LSC to maintain a stable collective economic income. Second, the renewal involved villagers' housing sites, in which negotiation was no longer between the government and the LSC collective organization, but included a large number of villagers. Third, renewal of old villages involved complex works such as moving villagers out and helping settle them elsewhere during renewal, and moving them back to the new sites. Fourth, as it was difficult to demolish all the houses of all the LSCs at the same time, the solutions to address the fragmentation problem in the second stage could not be adopted in this stage once again.

BH was the first LSC to experience negotiation with the government. According to the scheme proposed by the government, its remaining land could be registered by the government as state owned, administratively allocated to BH, and then transferred publicly by BH LSC as a package to a developer for integrated redevelopment. In the package, the developer should relocate villagers, demolish their houses, and construct collective properties for BH LSC and new homes for the displaced villagers. Land is cleared out for redevelopment, and government collected revenue from land transfer.

The floor area of the rebuilt property could be equal to that of the old factories. Alternatively, it could be equal to the amount needed to maintain a constant collective income flow seen before renewal. As to the redevelopment of villagers' houses, the room area of their previous government-registered housing could be fully compensated. Otherwise, the area of the housing would be considered illegal and compensated by cash at a discounted rate. During redevelopment, villagers had a choice between cash compensation and relocation to makeshift housing. After moving back to their new homes, the makeshift housing should be returned to developers. The government and BH LSC would invest in the redevelopment at this stage if the cost was too high to attract developers.

Although BH LSC agreed with the compensation for housing, it had its own plan related to the first two proposals. According to the BH plan, the government and BH should share the cost of developing 52.08 mu of non-residential land into properties of 138,900 m^2. The rest of the land, approximately 260.83 mu, was to be developed as a package. Moreover, the developer would provide villagers with both makeshift housing and case compensation. After the villagers moved into their new homes, the makeshift housing, with a floor area of 42,000 m^2, would be turned over to BH as a property asset. At that point, the rebuilt property area for BH would reach 180,900 m^2, far more than its original area of 135,900 m^2 and more than the 154,200 m^2 needed for a constant income flow.

The two proposals show fierce competition for floor area among the government, BH LSC, and the developer. Any developer that receives the package must pay two types of costs: one is revenue to the governments, the other is the cost to demolish old housing, place villagers in temporary housing, and reconstruct collective properties and new housing. The developer benefits from the floor area which he/she can freely transact in the market. The greater the compensation requested by BH LSC and villagers, the greater the costs to the developer. The developer will consequently require more floor area. However, the government has to ensure the bearing capacity of infrastructure and basic public facilities by controlling the total floor area of profitable land under the planned ceiling value. The other four LSCs have not yet begun the negotiation process, but it is certain they will follow the strategy adopted by the LSC of BH. Whether redevelopment will be implemented depends on how the government responds.

(3)   The key roles of government in land adjustment and transactions

In general, there were three roles that Nanhai local government played during the renewal process:

1)   In order to facilitate the renewal process, local government has to help achieve consensus among the LSCs and their members. Usually the government bore the early costs, and compensated the original tenants and paid rent to the LSCs. The local government then shared land rent differential by the LSCs which further incentivized bottom-up redevelopment.

2)   The government took strategies to assemble and adjust land to facilitate integrated redevelopment. In the second-stage renewal, the government demolished all factories of the LSCs, leveled the land, negotiated with the LSCs to reach agreement of redevelopment modes, and then pooled the land plots belonging to different LSCs to attract developers. In the third-stage renewal, the government made plans to provide the solution to address the fragmentation problem. On the one hand, roads were planned to run along existing territory boundaries of two adjacent LSCs when reallocating land plots, reducing the number of LSCs needing land adjustment. On the other hand, the plan also attempted to designate the same location, uses, and development intensity to the adjusted land plots. This made it possible to exchange land only according to land area. When the areas of land exchanged by two LSCs were different, the differentiated part could be transacted in cash. This simplified the complicated land adjustment problem and reduced transaction costs to a great extent.

3)   The government created different modes of land transactions according to different land uses and intentions of the LSCs. For old factories, land was leveled and withdrawn by the government and then publicly conveyed to developers. For the leveled land that the LSC wanted to develop on their own, the land was registered by the government to change landownership and then allocated to the LSC at no cost. For the land plots involving residential land, the land could be registered by the government to change the landownership, allocated to the LSC, and then publicly transferred to developers without demolishing the houses in advance.

Strategies of urban management of local government also worked well in the renewal of XB. The entire site was planned for the financial and business district of Nanhai, and the high-end urban functions brought about large land rent differentials. Also, the government successfully attracted several large developers, and these flagship projects rapidly advanced recognition of the markets as good for the future of the area. Last but not least, profit concessions of the government to the LSCs were also a key driving force of renewal.

(4)   Hold-out problem in redevelopment

Hold-out problems tend to exist during institutional transition. Transitional institutions are tentative and imperfect by nature, suggesting they are incapable to provide sufficient certainty to an emerging market. Institutional uncertainty usually leads to high transaction costs which makes it difficult to write a complete contract. When parties to a future transaction must make relationship-specific investments in advance and the specific form of an optimal transaction, such as time of delivery and quantity of units, is not yet determined with certainty, hold-out problems may arise (Williamson, 1979; Rogerson, 1992). Accordingly, the party that has made a prior commitment to the relationship could be "held out" by the other party for the value of that commitment, potentially leading to high economic costs.

The TOR program is essentially a change in the land rights institution. During the land rights change, village collectives' rights over land are strengthened and clarified. However, as a characteristic of transitional institutions, institutional uncertainty exists as part of the TOR. On the one hand, institutions frequently change. On the other hand, some specific rights over land between the villages and the government are still not clearly defined, such as the aforementioned land development rights. Under this context, it is understandable for village collectives and governments to write an incomplete contract for complicated redevelopment which could result in a hold-out problem.

In the renewal of XB village, the contract was incomplete in terms of regulations about revenue distribution among the stakeholders. First, there was no effectively executed control plan. The redevelopment office began to draw a plan in 2010, and the plan passed expert review in 2012. However, the government neither submitted the plan to the higher level planning authority for approval nor solicited public opinion. Without monitoring from the public and higher levels of government, maintenance of the environment through planning control was at the discretion of local governments. The plan was only known to the local government and seemed to be its instrument to negotiate with the LSCs, as the indices of the plan such as floor-area-ratio and building density were still negotiable. The boundary between individual revenue and public profit was unclear. Second, although the contract specified the proportion of revenue distribution between the two parties with respect to different types of land transactions, the distribution of overall land rents during the process was unclear.

As mentioned before, the third-stage renewal needed heavy expenditures with little return on investment, which is different from the second-stage renewal which generated substantial revenue. Money was immediately distributed after a land plot was conveyed in the second stage, and the money received by the LSC committee and local government from the second-stage renewal should fund the third-stage redevelopment according to the contract. However, it was unknown how much each party would pay. The profit-sharing regulation was ambiguous. Given the incomplete contract, the local governments' substantial *ex ante* investment and prior commitment to the LSCs finally resulted in its hold-out by the LSCs.

After the second-stage renewal, the intention of government to promote continued redevelopment and the incomplete contract provided opportunities for the LSCs to bargain for increased compensation. For instance, according to BH's earlier proposal, the total floor area acquired by BH was 259,200 m^2, including the rebuilt property and villagers' housing. The smallest amount of floor area required by the developer was estimated to be 269,200 m^2. The total floor area in the third-stage renewal was 587,400 m^2 in order to achieve balance among stakeholders, nearly close to the planned ceiling value of 600,000 m^2. The net cash income obtained by BH village, including both the LSC committee and the villagers, was estimated to be 489.3 million Yuan, and the net income that the local government received from BH's redevelopment was 460 million Yuan. The proportion of net cash income between BH and the local government was greater than 5:5. If considering the value added of new properties and housing relative to the old factories and housing, the proportion of net earnings could go as high as 8:2(Guo et al., 2017).

For the three LSCs of LN, ZB, and LB which sold most of their leveled land, the third-stage redevelopment would be much more difficult. Rough estimation, based on BH's experience, shows that if the government wants to implement the plan, it has to take out the total revenue received form the entire redevelopment of the three LSCs to compensate developers. Alternatively, the government has to relax the planning control and allow higher development density which will pose heavy pressure on the capacity of infrastructure, and finally the cost has to be borne by the public.

6.4 TOR in Panyu district: characteristics and spatial lock-in

6.4.1 Study area

The Panyu district is located in the southern periphery of Guangzhou city (Figure 6.10). It used to be a county-level city, but was merged into Guangzhou in 2000 and became an inner suburb district. It covers a land area of 529.94 km^2, including six township governments, ten street communities (*Jiedao* in Chinese) and 177 administrative villages within its jurisdiction at the end of 2012.In 2016, its GDP reached RMB175.4 billion Yuan, and its total permanent population was 1.64 million, consisting of 36.1% local residents with a household registration (*hukou*) while 63.9% were migrants (Panyu Statistics Bureau, 2012-2017).

6.4.2 City expansion and fragmented land use

Since 2000, Panyu has expanded rapidly, and its non-agricultural land grew from 134.45km^2 in 1999 to 217.3 km^{2}in 2012. In 2014, its total urban–rural

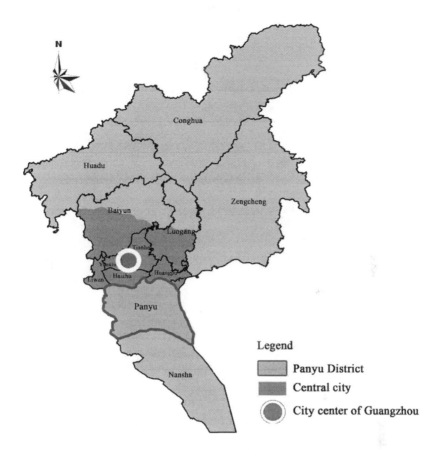

Figure 6.10 Location of Panyu District in Guangzhou city
Source: drawn by the authors

construction land accounted for 44.9% of all land. Due to the industrializa-
tion-led growth path, the manufacturing industry has been quite developed.
State-owned land and collective-owned land is spatially interwoven and
embedded as shown in Figure 6.11, and almost one-third of construction
land is rural collectively owned land.

6.4.2.1 State-owned and Collective-owned construction land in Panyu, 2013

As of 2014, a land area of 60.52 km² was listed as TOR patches, accounting for
29.82% of all construction land in Panyu (Figure 6.11). Among the land
patches, the percentage of collective land reached 64.51%. Table 6.8 presents
the composition of the TOR patches, showing that old factories took the largest
share of 58.84%, while old villages and neighborhoods accounted for 33.01%

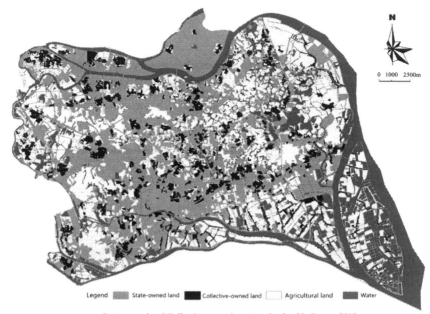

Legend ▨ State-owned land ■ Collective-owned land □ Agricultural land ▨ Water

State-owned and Collective-owned construction land in Panyu, 2013

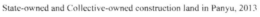

Legend ■ Old Neighborhoods Patches ■ Old Villages Patches ▨ Old Factories Patches

Figure 6.11 Spatial layout of TOR patches in Panyu
Source: Panyu Urban Renewal Bureau, 2015.

Table 6.8 Composition of TOR patches in Panyu district

Type	Land Area and Percentage (km²)	Property Structure (%)		Rights	Percentage of Land with Legal Title
Old neighbourhoods	4.93(8.15%)	State-owned	49.38%	88.5%	
		Collectively owned	50.62%		
Old villages	19.98(33.01%)	State-owned	2.03%	67.97%	
		Collectively owned	97.97%		
Old factories	35.61(58.84%)	State-owned	45.75%	57.57%	
		Collectively owned	54.25%		

Source: Panyu Urban Renewal Bureau (UPB), 2015

and 8.15%, respectively. Among the old factories land, around 19.32km² was collective industrial land, only 35.63% of which was legally developed.

LSCs gave rise to unique compartmentalized industrialization and fragmented urbanization (Zhu & Guo, 2015). After nearly 30 years of bottom-up industrialization led by autonomous villages, there are now 308 collective-owned industrial parks in Panyu. There were 1,367 production teams from 177 administrative villages who are the *de facto* owners of collective industrial parks. The average area of collective industrial land patch is less than 1.57ha, and there are 977 industrial land patches spatially dispersed over every village. The 2009 survey from the URB shows that over 60% of collective industrial land was developed without a land-use certificate, i.e., without official approval from the government. As a result, industrial land patches with formal and informal LURs were mixed, further deepening the degree and dimension of land-use fragmentation (Wang et al., 2018). Ambiguous and incomplete property rights over collective land help create a land development market where covert and disorderly competition for land rent differentials prevails (Tian & Zhu, 2013).

6.4.3 From informal renewal to formal redevelopment in Panyu: an overview of TOR practice

(1) Attempts to formalize informal renewal

Before 2009, urban renewal was spontaneously renovated without approval from Panyu government due to the ambiguity of property rights over collective land. Driven by the maximization of land rent capture, land users informally changed their industrial land into commercial or office use, and sometimes even commodity housing without legal property rights.

The lax governance in regard to informal redevelopment has suppressed formal development, as the latter requires huge transaction costs. Formalizing these informal developments involves complicated social and economic issues, entails land rent seeking and dissipation (Zhu, 2016). Due to the ambiguity and insecurity of collective land rights, the urban state is very cautious in formalizing informal development of collective land. According to the Guangzhou Urban Renewal Bureau, over 80% of collective-owned land property is still illegal or informal after the change of land use, as the properties were subject to little formal development control (Ceng et al, 2017).

According to the Panyu URB, the land area of informal renewal projects reached 310.76 ha by 2010, among which 72.66 ha were TOR patches (Figure 6.12). By the end of 2010, around 182 ha of collective land was informally renovated under the auspices of a "Creative Industry Park."[4] Informal redevelopment had been tacitly approved and supported by district level government, as it was expected to generate taxes and activate stock land assets. As a result, land rent has been captured by existing landholders.

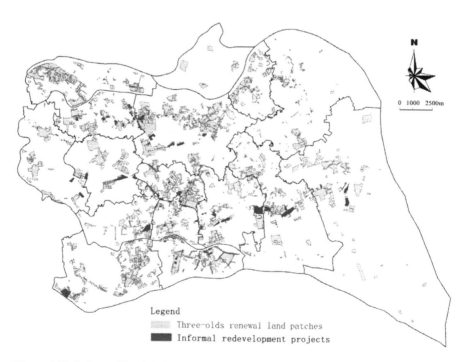

Figure 6.12 Informal land redevelopment in Panyu before 2010

Source: drawn by authors according to materials provided by Panyu URB

(2) Stages of TOR

Long-standing informal redevelopment has brought many problems such as low space quality, failure of government to capture land rent, and environmental problems. In 2009, Guangzhou issued "Guidelines to facilitate the work of TOR" (No.56 Document of Guangzhou Municipal Government) in order to formalize informal renewal and push forward the redevelopment process. Relevant regulations were also issued based along general guidelines. Overall, TOR can be divided into three phases in Panyu:

The first phase was the pilot period between 2009 and 2013. The policy was quite preferential for landholders and developers. If factory land use was changed to commercial or office use, the factory owners could be granted self-redevelopment rights and pay a land conveyance fee calculated on the benchmark premium differential between existing industrial use and new commercial or office use, and the upgrading of industry use was exempt of any land conveyance fees. If the industrial land was redeveloped for commodity housing, the redevelopment had to go through the auction, bidding, or listing process, and the land conveyance fee was shared between the government and factory owner. For old villages, they could be transformed into state-owned land with consent of the majority of the villagers. Otherwise, if collectives decided to retain their ownership, the collective land could be circulated to developers in terms of leasing, conveyance, sub-leasing, transfer, and mortgage to "comprehensive renovation" (Lin, 2015).In addition, collective land could be converted to state-owned through negotiation and LURs could be transferred to market entities for redevelopment. Villager relocation was negotiated under the principle of "One village, one policy" (*yicunyice*); however, village redevelopment was usually time-consuming and costly. The most challenging renewal was the old town, for there may be hundreds of property owners in an old neighborhood and the transaction cost to achieve consensus was extremely high. Motivated by the preferential policies to landholders, 19.48 km^2 of TOR land was approved to be redeveloped up to the end of 2012 in Guangzhou (Lai & Wu, 2013).

Considering the real estate market risk and the impact of property-led redevelopment on the city, the Guangzhou municipal government has slowed the pace of the TOR process since 2013. The second phase started from (or ran from, or was during) 2014 to 2015 and noted as the promulgation of No.20 Document in 2012. The new policy addressed the shortcomings of bottom-up scattered renewal and property-driven renewal, and the significance of comprehensive redevelopment. For instance, the government required that collective industrial land should be redeveloped together with the renewal of old villages in order to avoid the phenomena of "Meat eaten, bones left." However, this has dramatically lifted the threshold of renewal, and TOR almost stalled. By the end of 2015, the renewal of 39 projects, including 31 old factories, 4 old villages, and 4 old towns had been finished, accounting for 11.6% of all approved projects, and 97 projects had started construction at the city level (Table 6.9).

Table 6.9 TOR progress in Guangzhou in 2015

	Completed	*Under Construction*	*In Waiting List*
Old village	4 (9.8%)	15 (36.6%)	22 (53.7%)
Old factory (on state-owned land)	31 (11.4%)	75 (27.5%)	167 (61.2%)
Old town	4(18.2%)	7(31.8%)	11 (50%)
Total	39 (11.6%)	97 (28.9%)	200 (59.5%)

Source: Guangzhou Urban Renewal Comprehensive Report(Guangzhou URB, 2016)

In 2016, the Guangzhou government issued a set of renewal policies, the "Guangzhou Urban Renewal Regulation." According to this regulation, TOR approaches in Guangzhou can be divided into "Demolition and Reconstruction" and "Patchwork renovation."The former means site clearance and the transfer from collective land to state land. The patchwork renovation approach concerns the upgrading of human settlements and public facilities, and it is normally applied to old and dilapidated neighborhoods and historical areas. Compared with demolition and reconstruction, patchwork renovation does not incur a significant increase in floor space or pressure on transportation and public facilities, and is encouraged by the municipal government.

Since 2016, the TOR process has accelerated in Panyu, and 518.42 ha of stock land has been newly approved in the renewal list (Table 6.10). Three urban villages, *Shibi, Luobian,* and *Xinji,* were approved to implement *Demolition and reconstruction* in February 2018, and *Zuobian* village was approved to be renovated by the collective, another twelve old factory sites have also been approved. Since 2016, 39 dilapidated neighborhoods were selected by the municipal government for patchwork renovation. As of June 2018, around 37 km^2 of qualified TOR land had entered the redevelopment process, and 10.3 million m^2 of illegal building floor area had been demolished. Fixed investments generated from the TOR amounted to 23.42 billion Yuan, and the municipal government obtained a total of 12.85 billion Yuan in land conveyance fees.

6.4.4 Slow process of old town and village renewal

The TOR, so far, has played a vital role in promoting urban and rural redevelopment. A large amount of industrial land has been redeveloped, and constraints loosened upon collective land redevelopment, which has accelerated renovation of collective industrial parks. However, due to numerous property rights owners and complicated property rights over old towns and villages, their renewal has been quite slow. The ownership of old towns is complex and ambiguous, which includes public housing, work-unit

Table 6.10 Approved TOR projects in Panyu (2009–2018)

	Renewal Approach	*2009–2013*		*2016–2018*	
		Number	*Land Area(ha)*	*Number*	*Land Area(ha)*
Old village	Demolition and reconstruction	1	15.37	3	171.97
	Patchwork renovation	–	–	1	17.81
State-owned industrial land	Demolition and reconstruction	21	107.29	12	62.65
Collective industrial land	Demolition and reconstruction	1	7.35	4	86.12
	Patchwork renovation	–	–	2	17.46
Old neighborhoods	Patchwork renovation	–	–	39	97.41
Old town	Patchwork renovation	–	–	1	65
Total		23	130.01	62	518.42

Source: Panyu URB, 2016
Notes: There were no projects approved between 2014 and 2015

housing, and private houses. Only the local state has a capacity to assemble the fragmented ownership among these different residents (Li & Liu, 2018). Similarly, the old villages involve several hundred, and sometimes, even thousands of villagers, and the transaction cost to achieve consensus is very high, even with the support of rural collectives. Therefore, only four old town and four old village renewal projects have been completed in Guangzhou since 2009. In Panyu, up until June 2015, redevelopment plans for projects with a total land area of 130 ha had been approved by the Guangzhou municipal government. Among the approved projects, there was only one old village redevelopment project and no old town renewal projects, while the others were old factory renewal projects. Since 2016, only dilapidated neighborhood patchwork renovation projects have accelerated.

(1) An overall picture of dilapidated neighborhood renewal in Panyu

In Panyu, the 2016 Patchwork Renovation policy gave more leeway to villages in managing their renewal and refurbishment (Zhu, 2018). Between 2017 and 2018, around 42% of land area renewal projects consisted of dilapidated neighborhood renovation. In Panyu, neighborhood renovation aims at improving quality of public spaces and residential buildings. Patchwork renovation covers 59 items, and these are classified into specified items and optional items, with references to different investment sources.

Community public space items include fire equipment installation, municipal infrastructure maintenance, community roadway lighting, and illegal building demolition. Common parts of residential buildings include electricity, water and communication installations, external wall beautification, elevator equipment, roof greening, and energy conservation. Generally, government finance is the main source of dilapidated neighborhood renewal. In 2018, 66% of the annual urban renewal budget, nearly 231 million Yuan, went into dilapidated neighborhood renovation in Guangzhou.

In Panyu, between 2016 and 2018, 39 dilapidated neighborhood renovation projects were approved by the Guangzhou URB, with a total land area of 97.41ha and 1.59 million m^2 of floor space. Most of these projects were distributed in downtown and its nearby area of Panyu district, such as *Qiaonan* and *Shatou*. As of June of 2018, five neighborhoods had finished renovation, including external wall beautification and road improvement with a subsidy from the state.

(2)   Case of state-led renewal of old villages: the redevelopment of *Dongjiao* village

Dongjiao village has been the sole old village renewal project implemented in Guangzhou since 2009. The village is situated along the *Shiqiao* River, and is adjacent to the central area of Panyu. Since 2012, *Dongjiao* collective industrial land had been forbidden from being rented, and the *Dongjiao* collective income was around 6 million Yuan. This was much lower than the average level of collective income which was 16.36 million Yuan in2013. Villagers were eager to redevelop their industrial land to help grow their economy. With the opportunity of extending trunk roads and the building of public hospitals in *Dongjiao* village territory, redevelopment kicked off in 2013. The redevelopment covered a land area of 15.4 ha, involving 968 households (2,674 villagers), and land property rights were diversified, including collective land (housing and commercial/office), state-owned land (industrial, public hospital) and state reserved land.

The project was basically a joint redevelopment project among the state, village collective and private developer. In order to push the redevelopment process forward, local government acquired the industrial land and built relocated houses for villagers. The policy for compensation and resettlement was a one-to-one (in square meters) exchange (*Chaiyi buyi*) for any rural housing with legal titles. In other words, for every demolished property, new housing of an equal size could be claimed for resettlement (Lin, 2015). The average relocated floor space per household was 203 m^2 in *Dongjiao* village, and the maximum floor space per household was 280m^2, depending on the existing legal floor space. The total relocation floor area was 1.97 million m^2, and villagers preferred in-site relocation. The local government provided its reserve land for relocation, and invested substantially in the construction of public facilities.

Table 6.11 and Figure 6.13 show the land use change of *Dongjiao* village after the renewal. The share of public facilities and roads grew substantially, owing to an investment of around 1.197 billion Yuan from the Panyu government for the extension of district-level trunk roads, hospitals and high schools, and the relocation of villagers. Collective construction land decreased from 62.75% to 6.31% after land expropriation, while public facilities and spaces increased to 32.21%. After the renewal, the average FAR increased from 1.45 to 1.93, among which the FAR of commodity housing and collective property reached 3.88 and 1.78, respectively.

Table 6.11 Land-use change of *Dongjiao* village before and after redevelopment

Original Land-use Type	Land Area (ha) and its Percentage	New Land-use Type	Land Area(ha) and its Percentage
Collective commercial/office	2.97(19.27%)	Commercial/office	0.97(6.31%)
State-owned industrial use	1.48(9.60%)	–	–
Collective housing	6.70(43.48%)	Residential use	5.01(32.6%)
Public facilities	1.26(8.18%)	Public facilities	3.45 (22.58%)
Land for roads	1.70(11.03%)	Land for roads	4.06(26.42%)
–	–	Open space	1.48(9.63%)
Others	1.30(8.44%)	Others	0.38(2.47%)

Source: Redevelopment Plan of *Dongjiao* Village, 2013.

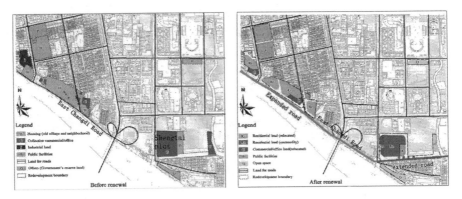

Figure 6.13 Land-use change of *Dongjiao* village before and after renewal.
Source: Panyu URB.

In this redevelopment project, the local government did not make both ends meet. The government was assumed to undertake receive 1.197 billion Yuan, but only received a land conveyance fee of 804.5 million Yuan from a collective property land auction. Also, the district government had invested 127 million Yuan for renovation and beautification of the surrounding area of *Dongjiao* village. The redevelopment of *Dongjiao* was not intended to generate profit for the government. In fact, the government had to give up land income in order to make the project financially viable (Wu, 2018). Sixty percent of the land conveyance fee from developers was distributed to the village collective, and the remaining 40% was required to compensate the relocation and infrastructure fee for the project by the district government.

Through the renewal, the *Dongjiao* village collective made much profit. The collective received 395 million Yuan cash for the compensation of collective property. In return for giving up collective land, the village collectives and villagers acquired houses with a floor space of 126,400m^2 and commercial property with a floor space of 18,100m^2 (Table 6.12). With a significant government investment, the *Dongjiao* renewal was completed. However, it was impossible to expand the *Dongjiao* renewal model to other old villages. Overall, the renewal of *Dongjiao* village seemed like a win-win situation, and all stakeholders, especially existing landholders and developers, have benefited from the renewal.

However, the public have been more like "outsiders" in the TOR process, and excluded from the decision-making process due to the lack of a public participation mechanism. For tenants in *Dongjiao* village, they had to move to other villages, or have had to bear higher rent caused by property-led high density redevelopment.

Table 6.12 Cost and benefit of *Dongjiao* village collective in the renewal

	Item	*Capital*
Cost	Building cost of collective property	45 million Yuan
	Land conveyance fee for collective property	59.35 million Yuan
	Building cost of kindergarten and other facilities	12.39million Yuan
Benefit	Land conveyance fee for *Shengtai* plot	482.7 million Yuan
	Cash compensation for displacement	29.44 million Yuan
	Collective property(state-owned land)	18,100m^2 of floor area
	Relocation space for villagers	126,400m^2 of relocation floor area
	Public facilities	3,044m^2 of floor area

Source: Panyu URB.

(3) Case of collective-owned industrial land: renewal of *Luoxi* industrial
park

Luoxi village is located in the west end of the *Luopu* Street Community in
the northwest corner of Panyu, with bridge access and is adjacent to
Guangzhou central city. From 2007 to 2013, the *Wuhusihai* Group (hereafter *Wuhu*) rented vacant collective land from *Luoxi* LSC for a 20-year
period, and transferred it to the International Aquaculture Trading Center
in 2009. However, only 20% of the space was successfully rented. In 2013,
Wuhu decided to transform the project into a commercial complex, called
Guangzhou Fisherman's Wharf (hereafter *Fisherman's Wharf*). The project
covered a land area of 62.6 ha, with state- and collective-owned land, and
legal and illegal land interwoven (Figure 6.14).

Fisherman's Wharf included 18 land patches, and the project was developed in three stages. In the first stage, six pieces of industrial land were
informally transformed into restaurant, creative workshop, SPA, and other

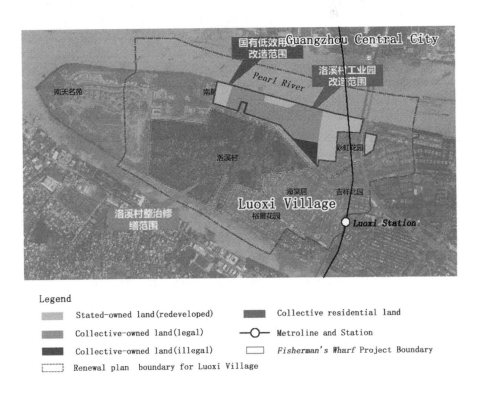

Legend

Stated-owned land(redeveloped)		Collective residential land	
Collective-owned land(legal)		Metroline and Station	
Collective-owned land(illegal)		*Fisherman's Wharf* Project Boundary	
Renewal plan boundary for Luoxi Village			

Figure 6.14 The status quo land property of *Fisherman's Wharf* project
Source: drawn by author based on material provided by Panyu URB, 2016

commercial uses or entrepreneurial incubation building. *Wuhu* invested 200 million Yuan to decorate or redesign existing buildings. In 2016, the second-stage renovation was formally initiated. A land-use plan for redevelopment designated the redevelopment areas and guided future land use. Before implementation, the Panyu URB organized a general renewal plan for the entire *Luoxi* village in February 2016, and designated a certain amount of land for infrastructure and open spaces. The main objective was to assemble collective industrial land and state-owned vacant plots, formalize illegal collective land, and convert some riverside vacant industrial land into green space. Land expropriation was also applied to secure public facilities and open space provision for the social good.

According to the renewal plan, 2.97 ha of state-owned land and 0.72 collective-owned warehouse land would turn into green space, and 30% of illegal collective land would be given to the local state. In return, the remaining 0.84ha of land was formalized and consolidated to a new land use. Land for infrastructure and sanitation (1.21 ha) and a kindergarten was provided. The third stage was upgrading the *Luoxi* old village through patchwork renovation. Table 6.13 shows the land-use change of the *Fisherman's Wharf* project after the renewal, and collective industrial land converted to high-rent commercial property through patchwork renovation at the expense of a slight decrease from 9.8 ha to 8.51 ha.

In terms of the input and output of this project, the total investment was around 1,000 million Yuan. The government assumed the cost of land expropriation and plan making, and the land conveyance fees were not enough to

Table 6.13 Land-use change of *Fisherman's Wharf* project

Before Renewal		After Renewal		
Classification	Land area (ha)	Classification	Land area (ha)	ratio
Collective-owned industrial land	9.8	Collective commercial property	7.95	12.7%
–	–	Public facilities	0.56	0.9%
Collective residential land	32.2	Collective residential land	32.2	51.4%
Agricultural and water land	16.6	Agricultural and water land	16.6	26.5%
State-owned factory and warehouse	4	Municipal infrastructure and road	1.52	2.4%
		Riverside green space	3.69	5.9%
		Government property	0.08	0.1%
Total	62.6	Total	62.6	100%

Source: author's calculation based on renewal plan

cover the cost. Government's profit was mainly derived from tax revenue and the beneficial social effect. Up to 2015, the project had generated 250 million Yuan in revenue, attracted 826 enterprises, and created around 6,000 jobs (Guangzhou URB, 2016). Developer *Wuhu* was the primary investor for the project, and needed to invest 535 million Yuan in renovating the plants, providing public facilities, and they expect to earn net rent of 640 million Yuan over the next 20 years after paying rent to the *Luoxi* collective. Renovation greatly stimulated rental growth, and the commercial rental price of *Fisherman's Wharf* rose to 100–200 Yuan/m^2/day in2015, much higher than the 18 Yuan/m^2/day of nearby collective property rent. For the *Luoxi* collective, they did not need to pay a land conveyance fee for land use change because the public facility land was contributed. The collective property rental income was expected to reach up to 376 million Yuan over 20 years, equivalent to 18.80 million Yuan every month, almost four times the 2012 level.

The *Fisherman's Wharf* project has achieved the goal of renovation, and collectives and developers have benefited from the renovation. The landscape has substantially changed after the renewal. Its success is greatly attributed to the good location, since *Luoxi* is the portal area of the Panyu district with metro stations nearby and close proximity to the central city. In general, village sites with better accessibility to transportation facilities and the city center are significantly more likely to be redeveloped than those located in less accessible areas (Lai & Zhang, 2016).

6.4.5 Property-led redevelopment dominated by state industrial land

(1)   An overall picture of state industrial land renewal in Panyu

As of June 2015, the redevelopment plans for 23 projects, including a land area of 130.01 ha, have been approved by the Guangzhou municipal government. Among the 23 projects, there was only one old village redevelopment project, while the other 22 were old factory renewal projects. Seven projects began construction, involving a land area of 52.09 ha, and the redevelopment of another 16 projects, involving a land area of 77.92 ha, have yet to begin construction. Among the 22 old factory renewal projects, there was only one collective-owned factory, while the other 21 factories were built on state-owned land.

The renewal of industrial land was facilitated by property-led redevelopment. All of the projects were planned to be reconstructed as commercial use, including a commercial center and office buildings or residential areas. The rent gap between industrial use and residential use was much larger than that between residential use and new property development in Panyu.

In terms of both land-use type and intensity, significant changes have or will have taken place in these renewal projects. The land use change of 22 industrial renewal projects show that the average Floor Area Ratio (FAR) would increase from 0.73 before renewal to 2.43 after renewal, and industrial land would be converted into commercial/office (48.89%), commodity housing (34.93%), roads (13.89%), and open space (2.29%) after the renewal (Table 6.14). According to the redevelopment plan for the 23 projects, there would be a total of 1,630,000 m² floor space in the newly added commercial/office projects and 1,230,000m² of new floor space of commodity housing. Property-led redevelopment posed huge challenges for the real estate market capacity of Panyu where the vacancy rate of commercial property ranged from 21% in 2013 to 10% in 2015, higher than average level in Guangzhou.[5] As all of these industrial land renewal involved demolition and reconstruction, the "bulldozer" approach to renewal also broke up social networks, and a spatial gentrification problem emerged.

Table 6.14 Land-use change of 22 industrial land redevelopment projects

Project	Land Use After Redevelopment	Land Area(ha)	FAR(status quo)	FAR After Redevelopment
A1	Commercial complex	7.35	1.0	2.84
A2	Shopping mall and tourism use	0.8	0.09	**3.11**
A3	Shopping mall, hotels, and enterprises headquarter	1.39	1.13	**3.01**
A4	Shopping mall and serviced apartment	2.68	1.25	2.89
A5	Commercial complex	15.10	0.53	2.65
A6	Office building and commercial center	8.06	0.57	2.77
A7	Business center	1.05	1.2	4.13
A8	Business center	1.49	1.06	2.32
A9	Commercial center, hotel, and office use	1.35	2.96	2.60
A10	Business complex	4.6	1.43	2.16
A11	Business center	1.2	1.4	1.97
A12	Commercial and office	21.28	0.39	2.15
A13	Creative enterprise zone	6.64	0.35	2.12
A14	Commodity housing	19.62	0.14	2.31
A15	Commercial and office	2.14	3.56	3.10
A16	Creative enterprise zone	2.71	2.7	2.72
A17	Business office and hotel	3.06	0.02	0.57
A18	Business hotel	2.27	1.92	2.57
A19	Commercial and office	8.7	1.3	≤2.0
A20	Business office and hotel	1.9	Unknown	0.6
A21	Business hotel	1.32	Unknown	3.09
A22	Commodity housing	3.10	Unknown	2.6

Source: Panyu URB, 2014

(2) Cases of state-owned industrial land renewal: Huamei and Zhujiang

Depending on the new land use, the renewal of state-owned industrial land can be classified as two types: self-redevelopment and profit-sharing. If the new use is for commercial or office, the land owner should pay a land conveyance fee, and they can form partnerships with outside capital for the renewal with a preferential land price. If the new use is for commodity housing, the land must go through the tender or auction process. LURs conveyance fees are shared between landholders (60%) and the local government (40%). Generally speaking, self-redevelopment is more prevalent than profit-sharing. From 2011 to 2018, 69.7% of factory owners chose the self-redevelopment model.

Taking the renewal of two factories, Huamei Wood Industry Co., Ltd. (hereafter *Huamei*) and Zhujiang Steel Pipe Limited (hereafter *Zhujiang*), as cases, we compared the cost of redevelopment under the traditional redevelopment approach and the TOR policy. According to the TOR policy, factory owners must pay the gap of benchmark premium differential between the existing industrial use and proposed commercial/office use in order to acquire the redevelopment rights. This gap was 281.51 million Yuan in the *Huamei* case and 425.53 million Yuan in the *Zhujiang* case.

However, if the traditional redevelopment approach had been adopted, the local government would have only paid 31.5 million Yuan to *Huamei* and 56.25 million Yuan to Zhujiang to acquire the LURs, plus limited compensation for the factory buildings and equipment, and they would have then sold the LURs on the open market. Based on the approved floor space of the renewal proposal, the new developer must pay 2,235 million Yuan for the *Huamei* plot and 4,007 million Yuan for the Zhujiang plot to the local government to acquire the land redevelopment rights (Source: Calculated according to the data provided by Panyu Land Banking Center in 2014). In other words, if the current land users acquire redevelopment rights in the open market, the *Huamei* owner will pay nearly eight times more, and the *Zhujiang* owner nine times more of LURs fees under the TOR policy in contrast to the traditional approach. The FAR of the *Zhujiang* plot will increase from 0.53 to 3.2, with a building coverage of 33.5% (Figure 6.15). Therefore, the existing industrial land users have been greatly motivated to initiate the redevelopment process.

Under the market-driven property-led renewal model, consideration of public benefits had not been on the top agenda of the local state. Among the 22 old factory renewal projects, only four projects with commodity housing developments were required to provide a floor space of 128,000m^2 for public facilities, while the remaining 18projects did not have to contribute any land for public use. None of the 22 projects were renewed for industry upgrading. Geographically, the 22 projects were scattered, focusing on areas with good locations and an active real estate market.

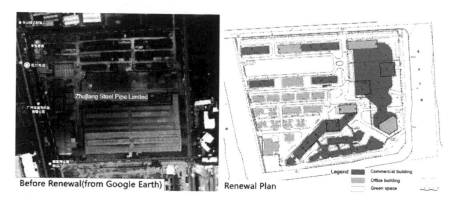

Figure 6.15 TOR plan of Zhujiang factory (left: before renewal; right: after renewal)
Source: Google Earth and Guangzhou Urban Planning Bureau, 2015

6.4.6 Why is collective land locked in the TOR of Panyu?

(1)   Complex procedures and high transaction costs to achieve consensus among villagers

Compared with the renewal of state-owned industrial land, the renewal of collective land has been much slower. The renewal procedures for old villages are complicated and involve many stakeholders: village collectives, villagers, developers, and tenants. Up to now, only *Dongjiao* village has been successfully renewed. Old village redevelopment involves two-rounds of public participation in order to achieve public consensus. First, a survey will be conducted if the area is going to be identified for renewal, and only after an 80% positive response rate has been achieved can the process move ahead. The second round survey focuses on the compensation scheme and the details of the redevelopment plan. If the compensation scheme does not achieve over 80% public consensus within 3 years after official approval, the redevelopment plan will be rejected. Moreover, TOR involves public benefits, collective benefits and individual benefits. Whether the stakeholders can achieve consensus is critical for the redevelopment process.

It is quite challenging to redevelop collective industrial land. According to the regulation, collective industrial land without a land-use certificate can be redeveloped on the condition that a village submits 30% of total land to local government and pay a penalty of 2 Yuan/m^2 of illegal land to the government. Between 2009 and 2017, there were, in total, 31 pieces of collective industrial land with a land area of 542.36 ha which were listed as TOR patches. However, collectives and the government did not reach an agreement on a profit distribution scheme. Only three projects (*Dongjiao, Luoxi, Longqi*) were implemented, and another five projects achieved the required

80% consensus rate and entered the redevelopment process. Twenty-three projects were stalemated by compensation bargaining among villagers, collectives, developers, and the government, and 96.47% of collective industrial parks have not begun the renewal process (Table 6.15).

(2) Path dependence of village's reliance on land and property leasing income

Due to the large amount of migrant workers in Panyu, most villagers expanded their houses to earn rent from the newly added space. Moreover, collective economic land was rented to outside enterprises, and villagers could obtain quantified shares from land rent and other sources (Tian & Zhu, 2013). In 2013, the total collective income in villages reached 2.895 billion Yuan. After deducting village reserved income and village management fees, the dividend left for collectives was 1.3 billion Yuan, as much as 3,465 Yuan per LSCs member. Around 65% of villages' land leasing income accounted for over 50% as a percentage of total village income (Table 6.16). Therefore, local villagers were not motivated as strongly as old factory owners, to initiate self-redevelopment, as they enjoyed permanent

Table 6.15 Progress of collective industrial park redevelopment

	Number of Industrial Parks	*Land Area of Industrial Parks (ha)*
In redevelopment process	8 (0.91%)	116.61 (6.04%)
Bargaining	23 (2.62%)	425.75 (22.04%)
Not initiated	846 (96.47%)	1389.64 (71.93%)

Source: calculated by author according materials provided by Panyu Urban Renewal Bureau.

Table 6.16 Contribution of land leasing income in total income of village in 2012

Land Leasing Income as a Percentage of Total Revenue	*Number of Village (%)*
75–100%	66(38%)
50–75%	47(27%)
25–50%	20(12%)
0–25%	40(23%)

Source: Panyu URB(2015)

dividends and fairly high rental income. Many villagers preferred an improvement in their environment without relocating, as wholesale demolition was infeasible in many villages, especially villages in those high rental demand areas. The collective land-use rights, owned by the villages, have constituted an effective holding power to bargain for institutional change to the collective land rights in favor of the rural collective. As a result, path dependent institutional change to the collective land rights has led to an entrenched collective (Zhu, 2018).

We collected the transaction records of collective-owned land and property from 2013 to the end of 2016, and the result reveals a large amount of collective asset circulation, as much as 144 records every month in Panyu district (Table 6.17). From 2010 to 2015, the total income of 177 villages rose from 2.4 billion to 3 billion Yuan (Panyu Statistics Bureau, 2012–2017), and the proportion of land and property leasing revenue grew from 43.23% to 50.97%. Since income from informal development had been substantial, villagers are cautious on bearing the risk of formal redevelopment of industrial land and prefer informal leasing in order to keep village revenue stable. Thus, there is a formal and informal competition between urban government and rural communities over collective industrial land use and development.

Table 6.17 Collective industrial land circulation records in Panyu (2013–2016)

	2013	2014	2015	2016
Sha Wan	27	223	196	174
Shi Lou	8	61	95	56
Hua Long	0	36	27	32
Shi Ji	22	72	165	89
Da Long	24	194	135	93
Nan Cun	0	106	128	220
Xin Zao	0	0	0	2
Da Shi	7	42	172	555
Luo Pu	0	76	85	52
Zhong Cun	0	100	57	30
Shibi Bi	0	80	43	40
Shi Qiao	6	305	317	189
Dong Huan	46	107	145	91
Sha Tou	6	159	135	88
Qiao Nan	0	151	99	94
Total	146	1,712	1,799	1,805

Source: collected from website http://nclqyj.panyu.gov.cn:8026/, accessed on 15/08/2018

Moreover, informal redevelopment of collective land has been in practice for a long time. Collective factories and warehouses are normally rented out to investors through land circulation from LSCs without planning permission. It is understood that spontaneous redevelopment by land users cannot bring high-quality redevelopment, and illegal property rights result in unequal access to credit. This has somewhat weakened the incentives of formal redevelopment (Lai et al., 2014).

(3)   High transaction costs to achieve census for renewal

The governance of villages has been fragmented and complicated in rural areas (Tian, 2008). Even within one village, the administrative village collective, natural-village collectives, and villagers have different expectations on redevelopment. It is challenging to achieve the consensus rate required in two rounds of public participation in old village redevelopment. Given the pressure from villagers, collectives are very cautious on bearing the risk of the redevelopment of village residential and industrial land, and thus, the renewal of collective land has been quite slow. Taking *Nancun* Town as example, 11 villages were included in the renewal plan owing to an arterial road passing through the town. This wholesale project was initiated as early as 2014. Only *Xinji* village agreed to adopt the approach of demolition and reconstruction, and after two years of negotiation, *Guantang* village agreed to patchwork renovation. The remaining night villages gave up on the plan for redevelopment. As of 2018, only the *Xin Ji* village renewal plan achieved the required consensus rate and its renewal plan was approved by the Guangzhou municipal government.

(4)   The absence of trust due to uncertainty caused by volatile redevelopment policies

TOR is basically a top-down process where the local government is active in initiating redevelopment projects. City governments have enough leeway to exercise discretionary power to execute the TOR policy, although the government no longer plays the dominant role in redevelopment by delegating self-redevelopment rights and a certain amount of land redevelopment profits to villages, it attaches many subsidiary conditions to constrain such rights. For instance, village collectives can only give up collective land ownership rather than choose to self-redevelop if they decide to convert industrial land use into commodity housing. Also, in order to redevelop collective land without a land-use certificate, village collectives must transfer about 30% of land to the local government.

Nevertheless, the policy design of TOR has been modified by the municipal government twice since 2009, revising some existing regulations and introducing several new constraints. For instance, since 2012, village collectives are no longer allowed to redevelop individual industrial land patches but instead, must take a holistic approach by redeveloping the industrial parks to which the individual land patches belong. Besides this change, profit sharing between local governments and villages when land is converted from collectively owned to state-owned used to be 30:70, but was changed to 40:60 in 2015.

The volatile redevelopment policies have given rise to an uncertainty of profit sharing and planning gains commitment, leading villages to become more hesitant to redevelop. If current institutional constraints are loosened, village collectives may gain more compensation or reduce their land losses. With an expectation of potential windfall profits, village collectives have tended to put off redevelopment in recent years, especially when village collectives find their vested interest will be unavoidably marred due to excess regulations.

During the long-period of bottom-up urbanization, the clan network played a more active role than the local government in village economic affairs. During the 1980s and 1990s, the pre-existing clan, kinship, and other personal ties between villagers and outside investors contributed in a large part to attracting investments and providing mutual trust for rural industrialization (Lin, 2001). For instance, a study of Hong Kong's production subcontracting activities conducted by Leung (1993) shows that many Hong Kong industrial enterprises in the rural area of the PRD were kinship factories owned by the relatives or friends of the respective Hong Kong investors. Respondents from Panyu confirm that some Hong Kong entrepreneurs preferred to locate their firm in their ancestral hometown.

Conflicts over historical land expropriation led to a distrust between the government and villagers. One of the major sources of conflict came from an "Economic Development Land (EDL)" (*jingji fazhan yongdi*) policy implemented in Guangzhou in 1993 which regulated that villages could receive an EDL of 10–20% of the acquired farmland area for collective economic development. Under some circumstances, village collectives found it hard to work with such quotas due to high administrative fees or strict planning restrictions (Wei & Yuan, 2007). As of 2015, EDL land quotas had not been realized in 58 village collectives, a problem which challenged the credibility of the local government.

6.5 Summary

TOR practices in Nanhai and Panyu reveal that granting self-redevelopment rights to current land users along with profit concessions of local state has pushed forward the redevelopment process of old factories. Nevertheless, the redevelopment has been strongly property-led and the comprehensive objectives of social and economic restructuring have not been achieved. Meanwhile, the TOR has failed in facilitating the renewal of collective-owned land and old towns. Problems such as an unclear risk-sharing plan, a lack of sustainable village economy, and an inefficient internal cooperative mechanism can result in the failure of the renewal policy. If these issues remain unaddressed, the endeavor of the current institutional change may be offset and the resulting prolonged redevelopment process may reduce the net profit of all involved parties. Therefore, further institutional change and more collaboration among various parties are required to overcome current barriers to spatial change.

The question of how to walk the path of further change remains a challenge, especially given that village collectives will have opportunistic behaviors confronted with volatile policy which was observed in Nanhai and Panyu. According to current redevelopment practices, there exists a possibility that further institutional change will provide more benefits to villages (Wang et al., 2018). As North (1991) and Libecap (1986) pointed out, the evolution of institutions largely depends on the bargaining power of influential actors, rather than the hypothetical wealth maximizing norm. Moreover, it is notable that interests of some related groups, such as migrant workers and current enterprises, rarely have their concerns addressed during the redevelopment process.

As some critical pioneering studies have noted, the efficiency gained from institutional change often comes with social costs (Lin, 2015; Guo et al., 2017). TOR does not address social equity or inequity and benefits for the public and tenants are often ignored. Establishing a cooperative renewal model which considers social and economic equity may be more appropriate and sustainable than simple profit concessions by the government. Further research on these issues must be conducted in order to provide guidance for future sustainable and inclusive redevelopment.

Notes

1 This percentage varies from city to city. It is 90% in Guangzhou city, and 80% in Shenzhen city.
2 In October 2011, The Department of Land and Resource of Guangdong province issued a notice about constructing a dynamically adjusted database of three-olds renewal land parcels. The already completed three-olds renewal projects are also included. Since issuing the notice, inclusion in the database is a prerequisite of renewal. The town/street governments submit to the higher-level government rankings of land parcels that meet certain three-olds renewal standards. The database is adjusted every six months. Based on the parcels included into the database, the Nanhai district government creates the TOR plan. According to the plan, the Nanhai district government identifies, approves and registers the renewal projects.
3 In LJ village, there is only one LSC managed by the administrative village committee. Most land in LJA belongs to LJ village. In order to reduce the number of parties in negotiation, the government requires each administrative village committee represent its subordinate village groups, in negotiation for redevelopment.
4 Since 2000, Guangzhou municipal government has issued a series of preferential policies to encourage the development of creative industry which allow the industrial land to be converted into the use for certain creative industry without need to pay for the land use change. In reality, sometimes these policies have been abused for the commercial use.
5 Source: www.colliers.com/zh-cn/china/insights#.VTMKsPSnU-0, accessed on 12/10/2015.

References

Barzel, Y. (1989). *Economic Analysis of Property Rights.* Cambridge: Cambridge University Press.

Benham, A., & Benham, L. (1997). Property Rights in Transition Economies: A Commentary on What Economists Know. In Nelson, J. M., Tiller, C., & Walker, L. (Eds.) *Transforming Post-Communist Political Economies,* Washington, DC: National Academy Presspp. 35–60.

Buitelaar, E. (2004). A Transaction-Cost Analysis of the Land Development Process. *Urban Studies,* 41(13):2539–2553.

Ceng, D., Wu, J., Huang, H. P., & Zhou, M. (2017). Exploration on "Micro – Transformation"of Old Factory in Guangzhou Based on Institutional Design: A Case Study of International Unit Creative Park. *Shanghai Urban Planning Review,* 5:45–50. (In Chinese).

Cheung, S. N. (1987). Economic Organization and Transaction Costs. In Durlauf, S. N., & Blume, L. E. (Eds.), *The New Palgrave: A Dictionary of Economics* (2nd *edition*), 55–58. London: Palgrave Macmillan.

Cheung, S. N. S. (1998). The Transaction Cost Paradigm. *Economic Inquiry,* 36(4):514–521.

Coase, R. H. (1960). The Problem of Social Cost. In Gopalakrishnan, C. (Ed.), *Classic Papers in Natural Resource Economics,* London: Palgrave Macmillan, 87–137.

Fischel, W. A. (1985). *The Economics of Zoning Laws: A Property Rights Approach to American Land Use Controls.* Baltimore, MD: The Johns Hopkins University Press.

Furubotn, E. G., & Pejovich, S. (1972). Property Rights and Economic Theory: A Survey of Recent Literature. *Journal of Economic Literature,* 10(4):1137–1162.

Furubotn, E. G., & Richter, R. (2000). *Institutions and Economic Theory: The Contribution of the New Institutional Economics.* Ann Arbor, MI: The University of Michigan Press.

Guangzhou Urban Renewal Bureau. (2016). Guangzhou Urban Renewal Comprehensive Report.

Guo, Y., Yang, X., & Yuan, Q. F. (2017). The Redevelopment of Peri-Urban Villages in the Context of Path-Dependent Land Institution Change and Its Impact on Chinese Inclusive Urbanization: The Case of Nanhai, China. *Cities,* 60(2):466–475.

Healey, P. (1992). An Institutional Model of the Development Process. *Journal of Property Research,* 9:33–44.

Lai, L. W. C. (1997). Property Rights Justifications for Planning and a Theory of Zoning. *Progress in Planning,* 48:161–246.

Lai, S., & Wu, J. (2013). Speed and Benefit: Guangzhou "Sanjiu" Redevelopment Strategies for New Urbanization. *Planners,* 29(5):36–41. (In Chinese).

Lai, Y., Peng, Y., Li, B., & Lin, Y. L. (2014). Industrial Land Development in Urban Villages in China: A Property Rights Perspective. *Habitat International,* 41:185–194.

Lai, Y. N., & Tang, B. (2016). Institutional Barriers to the Redevelopment of Urban Villages in China: A Transaction Cost Perspective. *Land Use Policy,* 58:482–490.

Lai, Y. N., & Zhang, X. L. (2016). Redevelopment of Industrial Sites in the Chinese "Villages in the City": An Empirical Study of Shenzhen. *Journal of Cleaner Production,* 134(5):70–77.

Leung, C. K. (1993). Personal Contacts, Subcontracting Linkages, and Development in the Hong Kong-Zhujiang Delta Region. *Annals of the Association of American Geographers,* 83(2):272–302.

Li, B., & Liu, C. Q. (2018). Emerging Selective Regimes in a Fragmented Authoritarian Environment: The "Three Old Redevelopment" Policy in Guangzhou, China from 2009 to 2014. *Urban Studies*, 55(7):1400–1419.

Libecap, G. D. (1986). Property Rights in Economic History: Implications for Research. *Explorations in Economic History*, 23(3):227–252.

Lin, G. C. S. (2001). Metropolitan Development in a Transitional Socialist Economy: Spatial Restructuring in the Pearl River Delta, China. *Urban Studies*, 38(3):383–406.

Lin, G. C. S. (2015). The Redevelopment of China's Construction Land: Practising Land Property Rights in Cities through Renewals. *The China Quarterly*, 224:865–887.

Ng, M. K., & Xu, J. (2000). Development Control in Post-Reform China: The Case of Liuhua Lake Park. *Cities*, 17(6):409–418.

NHBDPS (Nanhai Bureau of Development Planning and Statistics). (1998–2013). *Nanhai's Statistical Yearbook*. Foshan: Nanhai Press. (In Chinese).

North, D. C. (1991). Institutions. *The Journal of Economic Perspectives*, 5(1):97–112.

North, D. C. (2005). *Understanding the Process of Economic Change*. Princeton: Princeton University Press.

Panyu Statistics Bureau. (2012–2017). *Panyu's Statistical Yearbook*. Guangzhou: Codification Committee of Panyu Statistical Bureau. (In Chinese).

Rogerson, W. P. (1992). Contractual Solutions to the Hold-Up Problem. *The Review of Economic Studies*, 59(4):777–793.

Smith, N. (1979). Toward a Theory of Gentrification: A Back to the City Movement by Capital, Not People. *Journal of the American Planning Association*, 45(4):538–548.

Tian, L. (2008). The Chengzhongcun Land Market in China: Boon or Bane? A Perspective on Property Rights. *International Journal of Urban and Regional Research*, 32(2):282–304.

Tian, L., & Yao, Z. (2018). From State-Dominant to Bottom-Up Redevelopment: Can Institutional Change Facilitate Urban and Rural Redevelopment in China. *Cities*, 76:72–83.

Tian, L., & Zhu, J. M. (2013). Clarification of Collective Land Rights and Its Impact on Nonagricultural Land Use in the PRD of China: A Case of Shunde. *Cities*, 35:190–199.

Wang, B. Y., Tian, L., & Yao, Z. H. (2018). Institutional Uncertainty, Fragmented Urbanization and Spatial Lock-in of the Peri-Urban Area of China: A Case of Industrial Land Redevelopment in Panyu. *Land Use Policy*, 72:241–249.

Webster, C. (1998). Public Choice, Pigouvian and Coasian Planning Theory. *Urban Studies*, 35(1):53–75.

Wei, L. H., & Yuan, Q. F. (2007). How to Construct the City and Promote Urban Development on the Collective Land? A Case of Nanhai District in Foshan City. *Urban Planning Forum*, 3:61–65. (In Chinese).

Williamson, O. E. (1979). Transactions-Cost Economics: The Governance of Contractual Relations. *Journal of Law and Economics*, 22(2):233–262.

Wu, F. L. (2018). Planning Centrality, Market Instruments: Governing Chinese Urban Transformation under State Entrepreneurialism. *Urban Studies*, 55(7):1383–1399.

Ye, L. (2011). Urban Regeneration in China: Policy, Development, and Issues. *Local Economy: The Journal of the Local Economy Policy Unit*, 26(5):337–347.

Yuan, Q. F., Qian, T. L., & Guo, Y. (2015). Reconstructing Social Capital to Promote Urban Renewal: A Case Study of Lianjiao Area, Nanhai. *City Planning Review*, 39(09):64–73. (In Chinese).

Zhu, J. M. (2002). Urban Development under Ambiguous Property Rights: A Case of China's Transition Economy. *International Journal of Urban and Regional Research*, 26(1):41–57.

Zhu, J. M. (2016). The Impact of Land Rent Seeking and Dissipation during Institutional Transition on China's Urbanization. *Urban Affairs Review*, 53(4):689–717.

Zhu, J. M. (2018). Path-Dependent Institutional Change to Collective Land Rights: The Collective Entrenched in Urbanizing Guangzhou. *Journal of Urban Affairs*, 40, 1–14.

Zhu, J. M., & Guo, Y. (2015). Rural Development Led by Autonomous Village Land Cooperatives: Its Impact on Sustainable China's Urbanisation in High-Density Regions. *Urban Studies*, 52(8):1395–1413.

Zhu, J. M., & Hu, T. T. (2009). Disordered Land-Rent Competition in China's Peri-Urbanization: Case Study of Beiqijia Township, Beijing. *Environment and Planning A*, 41(7):1629–1646.

7 Towards compact and integrated urban–rural development in China

Peri-urban areas have been the most dynamic regions in China. While the economy of peri-urban areas has been vigorous, its social and environmental problems are worthy of attention. Land-use fragmentation is one of the most serious problems China is facing, and there are fundamental historical, economic, and institutional reasons for it. Historically, the TVEs burgeoning in the 1980s and early 1990s formed a spatial framework which has not yet been transformed in the peri-urban areas. Economically, constrained by a lack of human capital and low environmental quality, technology-oriented industries have had few chances to develop in peri-urban areas (Tian & Liang, 2013). Therefore, peri-urban areas have had to rely heavily on the manufacturing industry. Institutionally, formal and informal institutions coexist in the peri-urban areas. Loose planning regulation for collective land is one of key reasons for fragmented and scattered industrial use. This chapter summarizes the findings from the case study and explores the policy implications to achieve the goals of sustainable and integrated urban–rural development in the peri-urban areas as seen from the perspectives of economic development, spatial change, and governance change.

7.1 Transition from village-initiated industrialization to township/municipality coordinated development

7.1.1 Village-initiated industrialization and its impact on urbanization

As the basic unit of China's rural society, villages play an essential role in shaping collectivity which is beneficial for cooperation between village members. However, rural development is hampered by subsistence farming because of a deficient amount of arable land and the growing peasant population (Zhu, 2017). Currently, most village-initiated industries in peri-urban areas have a tight connection with the local government and some have adopted new forms of "industrial clusters" (Naughton, 2007; Huang, 2008). However, as summarized from the case studies in Chapter 4, the development of village-initiated industrialization in peri-urban China heavily relied on the manufacturing industry, while technology-oriented industries lacked a suitable environment in which they could prosper (Tian, 2015). Manufacturing often accounts for 60–70% of gross domestic product (GDP), and sometimes even higher (Tian, 2015).

Despite the monotony of the industrial structure, the development of village-initiated industries in peri-urban areas has significant limitations of unsustainability and fragmentation. On the one hand, subject to local institutional uncertainty, village-initiated industries tend to pursue short-term benefits more than long-term sustainable growth, neglecting industrial upgrading and regional joint developments. On the other hand, with a high population density and small-scale village autonomy, village-initiated industries developed independently, lacking divisions in the labor force, and a lack of overall planning, resulting in fragmented industrial sites and urbanization which generated dreadful problems of industrial sprawl and compromised industrial productivity (Zhu, 2017).

In general, village-initiated industrialization has created many job opportunities and attracted many floating workers. However, it has caused many problems which require the intervention of the upper level government. It is necessary for the peri-urban areas to coordinate economic growth at upper level government such as township or county government. The policies which can balance the interests of individuals, village collectives, and the public should be designed to enhance the production efficiency and enlarge local economies. Industrial development should evolve spatially from scattered village factories to concentrated industrial zones to make use of favorable industrial policies and infrastructure and attain sustainable development (Zhu, 2017).

7.1.2 Local attempts to coordinate urban and rural development

When looking at the cases of peri-urban areas, we find that Kunshan serves as a model of integration of rural villages into urban communities with a county/township coordinated development, and Nanhai acts as bottom-up industrialization-dominated growth example. Usually, land-use fragmentation and environmental problems are more serious in the latter. Since the late 1990s, when the TVEs were dismantled, Kunshan actively pursued agglomeration of industrial land, and these industrial parks have been mainly managed by the township and county government. Villages were no longer involved in economic growth, leading to integrated urban built-up areas. Moreover, the Kunshan municipal government adopted a fiscal transfer policy to mitigate income disparity among villages. The amount of fiscal transfers reached 34.7% of the total annual village collective revenues in 2012, and this has greatly reduced the incentives of villages to develop manufacturing industry within their own boundaries (Zhu, 2017).

Daxing district is another peri-urban area of Beijing, and has 14 townships under its jurisdiction. It covers a land area of 1,036 km^2, and has a population of 1.56 million. The proportion of primary, secondary, and tertiary industries was 4.29%:41.06%:54.65% in 2015. Since the 1980s, collective industry has been developed in the Townships within the Daxing district. In the 1990s, the Beijing municipal government issued policies to encourage the development of village industry courtyards (*Gongyedayuan*). Taking *Xihongmen* Township as an example, 27 village collectives established 27 industry courtyards and around one-third of its land, 10km^2, was made up of industrial courtyards (www.bjdx.gov.cn/jrdx/dxxw/mtbd/2335564.html, accessed on 9/19/2018). The scattered industry courtyards caused many environmental problems, such as pollution, traffic congestion, and low space quality.

In 2015, Daxing was listed as one of 33 pilot areas for establishing a market of collective construction land, and collective construction land was then allowed to be transferred in the open market except for use as commodity housing. In 2016, *Xihongmen* Township put forward an approach of "Township coordinated development," and set up *Shengshi* Assets Joint Venture Company (hereafter *Shengshi*) with 27 village collectives as stakeholders. *Shengshi* accumulated industrial land of 10 km^2 from villages, and demolished all factories. Eight square kilometers of land was used for an urban green corridor, and the other 2 km^2 was developed for commercial and office use. The dividend would be allocated to villagers based on their stakes, and was guaranteed an annual growth rate of 5%. Through the township coordinated development approach, the land use of *Xihongmen* became more compact and productive and the property rights over collective land was defined through negotiation among village collectives and villagers. With the land readjustment, land leasing income significantly increased, and the average household income of villagers increased 5,000 Yuan in 2017. Recently, Daxing district established other township coordinated companies to push forward land readjustment in other townships.

7.2 Transition of spatial growth patterns from rural fragmentation to compact urbanization

7.2.1 Rural fragmentation and significance of compact urbanization

The inefficient land-use pattern in peri-urban China has been commonly criticized (Tian, 2015; Zhu & Guo, 2014), and its negative impacts on sustainable development has been evidenced. Therefore, new planning policies were proposed and top-down industrial clusters have been created to develop a more concentrated land-use pattern. However, the fragmented spatial structure of peri-urban land has been hard to alter under a potential "lock-in" effect due to path dependence and high transaction costs of redevelopment. Fragmented land-use patterns would not only cause disorder in the spatial structure, but also reduce the efficiency of economic and ecological utilization to a certain extent. Although high-density land development in certain socioeconomic circumstances may have the risk of causing adverse phenomenon such as crime, vandalism, and social irresponsibility

(Fuerst & Petty, 1991; Gordon & Richardson, 1997), the low-density decentralized form of land use, considered as urban sprawl, has been defined as "uncoordinated growth" limited to the level of individual development instead of the sustainability of the region (Batty et al., 2003).

Curbing urban sprawl not only helps conserve the natural and agricultural landscape and retain land for future generations, but also shortens the daily commuting distances and reduces energy consumption and emissions of greenhouse gases (Elkin et al., 1991; Ewing, 1997; Cervero, 1998). Although it has been proposed that sustainable urbanization attributed more to the process of city building than the physical spatial form, compact urbanization is widely considered to be one of the most effective ways to pursue sustainable development in urbanized areas (Jenks et al., 1996; Jabareen, 2006). Containing urban sprawl and developing compact spatial structure in peri-urban areas are more conducive to sustainable urbanization. For example, the compact land-use mode helps save additional open space and green landscape from limited land supply, contributing to biodiversity within cities (Swanwick et al., 2003).

Recently, the shift from rural fragmentation to urban–rural integrity in land-use patterns has been a major trend in peri-urban areas. The agglomeration of industrial land helps increase productivity, upgrade industrial structure, and improve economic scale efficiency, which is also beneficial for effective provision of infrastructure and increasing rural income. Since the industrial restructuring in the late 1990s, collectively owned factories have been gradually replaced by private firms which tend to locate in planned industrial zones close to high-quality infrastructure. The tendency of spatial agglomeration helps enhance environmental integrity and promotes compact land-use patterns with less farmland consumed, indicating the movement of urbanization in peri-urban areas towards a more sustainable path (Zhu, 2017).

A compact spatial structure of urbanization is even more crucial in some Asian cities because of the high population density and scarce land resources. High-density and concentrated development improves the efficiency of land use in peri-urban areas. This makes governments available to provide utilities such as public transport and infrastructure in a more effective mode, support active social interactions, and promote social equity (Burton, 2000). In China, compact urbanization is a necessity rather than a choice in most areas due to the high population density and land scarcity. At present, the scale of the migrating population from rural areas to urban areas in China is expected to continue expanding. On the one hand, it is of great importance to enlarge the space capacity of existing urban areas through making good use of the stock of urban land so as to accommodate the rapid economic growth as well as the demand of urban newcomers' need for infrastructure (Zhu, 2017). On the other hand, for peri-urban areas that are still in the process of urbanization with unfixed land-use patterns, it is crucial to consciously change the fragmented spatial characteristics into intensive and compact land-use forms so as to provide a spatial framework for sustainable social and economic development in these regions.

7.2.2 Establishment of an integrated planning and management system

Land-use planning is essential for a compact and efficient spatial structure in order to promote orderly and sustainable development in peri-urban areas in China. Integrated planning of urban and rural land use is necessary to coordinate regional land development and guide the construction of a rational and efficient land-use spatial structure. Moreover, supporting policies, such as land-use management, finance, and taxation, should be designed for the implementation of land-use planning.

The stakeholders involved in the land-use planning of peri-urban areas are more diverse than that in well-developed cities or purely rural areas. The existing urban or rural planning system, however, is not sufficient to address the problems of peri-urban areas and achieve the goal of long-term sustainable urbanization (Tian& Zhu, 2013). Integrated urban–rural planning can be an effective tool to coordinate the overall regional space coordination.

In China, Chengdu municipality provides a successful example of urban–rural coordinated planning and development. It is located in the south-west region of China, and was listed as the first national pilot city in coordinating urban–rural integrated development in 2007. Chengdu put forward a model of "Planning for the whole domain" (*Quanyuguihua*), and 19 districts/counties, including 12,400 km^{2}of land area, were considered as an integrated part in the comprehensive planning (Figure 7.1). It was divided into five levels from municipality to rural community, and covered all land area within the municipal boundary.

The planning strategy of Chengdu was to realize three concentrations: concentrating industry in strategic function zones, concentrating land into big farms, and concentrating farmers into denser new-type rural communities. The Chengdu plan also included six integrations, integrated urban and rural planning, industrial development, markets, infrastructure, public services, and management. These have played a significant role in Chengdu's model for solving the fundamental Chinese problems concerning agriculture, the countryside, and farmers. They have reduced the gap between urban and rural areas and promoted urban–rural integration through increasing equality of public services, alleviating rural poverty, implementing rural development and environmental construction projects, new rural community construction, and policies to make it possible for rural and urban residents to enjoy the benefits of modernization together. From 2000 to 2016, the urban–rural income disparity in Chengdu decreased from 2.61 to 1.93, while the national average was 2.72 in 2016.

The key issue of this coordinated development was to break through the bottleneck of urban and rural elements' isolation, in particular, the land resource. In 2008, Chengdu initiated the reform of collective land market, and established a collective land exchange market. Through progressively making policy for transferring rural construction land-use rights, Chengdu set up a preliminary urban–rural integrated land market.

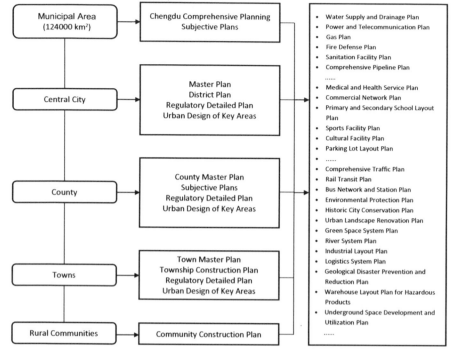

Figure 7.1 Urban–rural plan-making system of Chengdu
Source: Edited according toZhao and Zhu (2009)

7.3 Transition of governance model from rural autonomy to urban–rural integrated governance

In the process of urbanization and economic development in peri-urban China, bottom-up autonomy driven mainly by market-mechanism and top-down government integration dominated by central governance have always significantly influenced the regional development pattern.

7.3.1 Rural autonomy in the urbanization process

Rural communities applying a traditional social organizational mode has a long history in China. Village members are mostly bound by kinship and lineage, and thus the composition of the rural communities are involuntary and place-based (Zhu, 2017). In spite of the internal equality and autonomy, villages were usually not open to non-members, having few outside immigrants (Gao, 1999). Due to the homogenization of the population and the stability of social composition, social organizations in Chinese villages were usually autonomous, forming a self-contained socioeconomic unit based on clans and lineage.

Before the 1949 revolution, solidarity and coordination formed within rural communities under the autonomous management by the village (Skinner, 1971; Duara, 1988). Villages, considered as self-governing bodies (Gao, 1999), took responsibility for their own infrastructure and social services such as health, education, and elderly welfare (Wong, 1997; Tsai, 2007). After the 1949 revolution and the People's Commune Movement, the villages became the basic collective economic organizations in rural China.

Over time, however, the continued growth of rural population made land resources increasingly scarce. In the face of the opportunities of economic restructuring and the land rent differential under the dualistic system of urban–rural land management, some villages gradually entered the process of urbanization. However, the increase in land rent was not evenly distributed across rural areas as well-located villages attained more benefits from the land rent differential than other villages. In the transition to the non-agricultural economy, inequalities among different rural areas were created relying on land rent income provided by the urbanization progress (Shi, 2000). Under this circumstance, depending solely on rural autonomy had the risk of creating spatial inequality in peri-urban development (Zhu, 2017).

Urbanization tends to create social segregation, suburban sprawl, and other social problems if it occurs without the appropriate coordination of top-down planning (Duany et al., 2000; Roy & Alsayyad, 2004). Thus, the tensions between bottom-up rural autonomy and top-down urban government integration tend to rise during the urbanization of these peri-urban areas and have had a significant impact on local economic development and industrial transformation.

7.3.2 Transition of the governance model

Oi (1995) argued that the Developmental State is between the two extremes, the minimum intervention type or the central planning type of governance. Although the minimalist state of governance represented by the United States is highly respected in most western countries, the practice in some countries shows that the intervention of government can also play a positive role in regional economic development. In some rapidly growing Asian regions, it is generally accepted that governance has acted as an engine in regard to local economic development. In China, the role of government was not diminished during the transition to a market economy, but redefined (Tian & Zhu, 2013).

According to the case studies in Chapter 3, the different degree of correlation between the government and the market exerts different influences on the development of peri-urban areas in BTH, YRD, and PRD. At the beginning of reform and opening-up, village-initiated development under the rural autonomy was a creative mode that worked effectively in many rural areas. However, in the long-term perspective, as rural autonomy was confined within the boundaries of the homogeneous village (Tian & Zhu, 2013), a series of problems such as scattered industrialization and fragmented urbanization emerged. This was a result of the informal land development dominated by autonomous villages, leading to unsustainable regional development (Pei, 2005). In other words, informal systems may be efficient during a period of rapid social and economic transition, providing certainty and welfare for homogeneous communities in the process of developing non-agricultural economies.

Nevertheless, as urbanization is a dynamic and heterogeneous process, it is difficult for informal systems to deal with complex economic and social changes (World Bank, 2002). Top-down coordination can supplement the drawbacks of rural autonomy so as to avoid uncertainty and risk in regional development. The integrity of urban governance helps to reduce transaction costs so as to make the development process more efficient (Webster, 1998), and this is particularly important in peri-urban areas due to the ambiguity of collective land property rights and mixture of state-owned and collective land.

The township is a key spatial and administration unit in peri-urban areas. The governance boundaries of townships are relatively flexible and thus townships are seldom considered as a stable political entity (Zhu, 2017). However, they take on an irreplaceable coordinating and integrating role for promoting the healthy development and sustainable urbanization in peri-urban areas by providing rural areas with services and assistance so as to accelerate rural development, as well as passing the upper government's guiding policies downward to ensure regional coordinated development.

In order to form an urban–rural integrated governance, the traditional plan-making model, with striking drawbacks of separating urban and rural areas into two systems, cutting the link of vertical and horizontal bureaucracies as well as focusing only on technical design (Ye & LeGates, 2013), must be replaced by a new integrated plan-making and approval system. A holistic, comprehensive plan should be made for the entire metropolitan area across different geographic regions with long-term visions of its administrative, economic, and spatial structure reform. The public policy nature of planning should be emphasized to establish scientific urban–rural integrated planning and a corresponding plan evaluating and monitoring mechanism. The spatial layout structure should break the boundaries of urban and rural regions so as to improve land-use efficiency and intensity in the overall urban–rural integrated perspective, integrating economic and social needs with physical planning.

7.4 Conclusion

Rapid urban growth and the consequent land use changes have drawn much attention in China. It can be envisaged that in the near future, more people will be added to the peri-urban areas of China. Many issues facing peri-urban areas need to be handled from a wider spatial perspective (Webster & Muller, 2002). With a series of social and environmental problems caused by peri-urbanization, adjusting the urbanization strategy has been on the agenda of the Chinese government. In March 2014, the Chinese central government released its first official urbanization plan, the *National New-style Urbanization Plan* (2014–2020), among which the two most important issues were pushing forward the citizenization of migrant workers and making land use more compact and sustainable. In order to achieve these goals, institutional reform in collective land is necessary, and this requires more in-depth and further research from even wider perspectives.

References

Batty, M., Besussi, E., & Chin, N. (2003). Traffic, Urban Growth and Suburban Sprawl. *Working Papers Series*, Paper 70, UCL Centre for Advanced Spatial Analysis.

Burton, E. (2000). The Potential of the Compact City for Promoting Social Equity, In Jenks, M., Burton, E., & Williams, K. (Eds.) *The Compact City – A Sustainable Urban Form*, London: E&FN Spon, 19–29.

Cervero, R. (1998). *The Transit Metropolis: A Global Inquiry.* Washington, DC: Island Press.

Duany, A., Plater-Zyberk, E., & Speck, J. (2000). *Suburban Nation: The Rise of Sprawl and the Decline of the American Dream.* New York: North Point Press.

Duara, P. (1988). *Culture, Power, and the State: Rural North China, 1900–1942.* Stanford, CA: Stanford University Press.

Elkin, T., McLaren, D., & Hillman, M. (1991). *Reviving the City: Towards Sustainable Urban Development.* London: Friends of the Earth.

Ewing, R. (1997). Is Los Angeles-Style Sprawl Desirable?*Journal of the American Planning Association*, 63(1):107–126.

Fuerst, J. S., & Petty, R. (1991). High-Rise Housing for Low-Income Families. *Public Interest*, 103:118–131.

Gao, M. C. F. (1999). *Village: A Portrait of Rural Life in Modern China.* Hong Kong: Hong Kong University Press.

Gordon, P., & Richardson, H. W. (1997). Are Compact Cities a Desirable Planning Goal?*Journal of the American Planning Association*, 63(1):95–106.

Huang, Y. (2008). *Capitalism with Chinese Characteristics.* Cambridge: Cambridge University Press.

Jabareen, Y. R. (2006). Sustainable Urban Forms: Their Typologies, Models, and Concepts. *Journal of Planning Education and Research*, 26(1):38–52.

Jenks, M., Burton, E., & Williams, K. (1996). *The Compact City: A Sustainable Urban Form.* London: E & FN Spon.

Naughton, B. (2007). *The Chinese Economy: Transitions and Growth.* Cambridge, MA: The MIT Press.

Oi, J. C. (1995). The Role of the Local State in China's Transitional Economy. *The China Quarterly*, 144:1132–1149.

Pei, X. L. (2005). Collective Landownership and Its Role in Rural Industrialization. *Chinese Rural Studies*, 1:218–250. (In Chinese).

Roy, A., & Alsayyad, N. (2004). *Urban Informality: Transnational Perspectives from the Middle East, Latin America, and South Asia.* London: Lexington Books.

Shi, J. S. (2000). Rural Community Land Shareholding Cooperative System: Review and Prospect. *China Rural Economics*, 1:63–67. (In Chinese).

Skinner, G. W. (1971). Chinese Peasants and the Closed Community: An Open and Shut Case. *Comparative Studies in Society and History*, 13(3):270–281.

Swanwick, C., Dunnett, N., & Woolley, H. (2003). Nature, Role and Value of Green Space in Towns and Cities: An Overview. *Built Environment*, 29(2):94–106.

Tian, L. (2015). Land Use Dynamics Driven by Rural Industrialization and Land Finance in the Peri-Urban Areas of China: The Examples of Jiangyin and Shunde. *Land Use Policy*, 2015(45), 117–127.

Tian, L., & Liang, Y. L. (2013). The Industrialization and Land Use in Peri-Urban Areas: An Analysis Based on the Development of Three Top 100 County Economies in Three Regions. *Urban Planning Forum*, 34(5):30–37.

Tian, L., & Zhu, J. M. (2013). Clarification of Collective Land Rights and Its Impact on Nonagricultural Land Use in the Pearl River Delta of China: A Case of Shunde. *Cities*, 35:190–199.

Tsai, L. (2007). Solidary Groups, Informal Accountability, and Local Public Goods Provision in Rural China. *American Political Science Review*, 101(2):355–372.

Webster, C. (1998). Public Choice, Pigouvian and Coasian Planning Theory. *Urban Studies*, 35(1):53–75.

Webster, D., & Muller, L. (2002). *Challenges of Peri-Urbanization in the Lower Yangtze Region: The Case of the Hangzhou-Ningbo Corridor*. Working paper, Asia/Pacific Research Center, Stanford, CA: Stanford University.

Wong, C. (1997). *Rural Public Finance: Financing Local Government in the People's Republic of China*. Hong Kong: Oxford University Press.

World Bank. (2002). *World Development Report 2002: Building Institutions for Markets*. New York: Oxford University Press.

Ye, Y., & LeGates, R. (2013). *Coordinating Urban and Rural Development in China: Learning from Chengdu*. London: Edward Elgar.

Zhao, G., & Zhu, Z. (2009). Idea and Practice of Integrated Urban and Rural Planning in Chengdu. *Urban Planning Forum*, 6:12–17. (In Chinese).

Zhu, J. M. (2017). Making Urbanisation Compact and Equal: Integrating Rural Villages into Urban Communities in Kunshan, China. *Urban Studies*, 54(10):2268–2284.

Zhu, J. M., & Guo, Y. (2014).Fragmented Peri-Urbanisation Led by Autonomous Village Development under Informal Institution in High-Density Regions: The Case of Nanhai, China. *Urban Studies*, 51(6):1120–1145.

Index

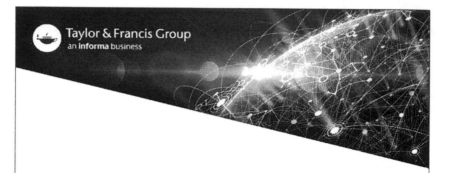